Origami Polyhedra Design

Origami Polyhedra Design

John Montroll

A K Peters, Ltd.
Natick, Massachusetts

Editorial, Sales, and Customer Service Office

A K Peters, Ltd.
5 Commonwealth Road, Suite 2C
Natick, MA 01760
www.akpeters.com

Library of Congress Cataloging-in-Publication Data

Montroll, John.
  Origami polyhedra design / John Montroll.
      p. cm.
  ISBN 978-1-56881-458-2 (alk. paper)
  1. Origami. 2. Polyhedra in art. 3. Polyhedra--Models. I. Title.
  TT870.M567 2009
  736'.98--dc22

                                                          2008053787

Cover photograph: Models folded by John Montroll and John Szinger. Photograph by Gabor Demjen.

Printed in the United States of America

13 12 11 10 09   10 9 8 7 6 5 4 3 2 1

To Himanshu

Contents

Simple	☆
Intermediate	☆☆
Complex	☆☆☆
Very Complex	☆☆☆☆

Symbols

Lines

— — — — — — — — — Valley fold, fold in front.

—·—·—·—·—·— Mountain fold, fold behind.

———————— Crease line.

··· X-ray or guide line.

Arrows

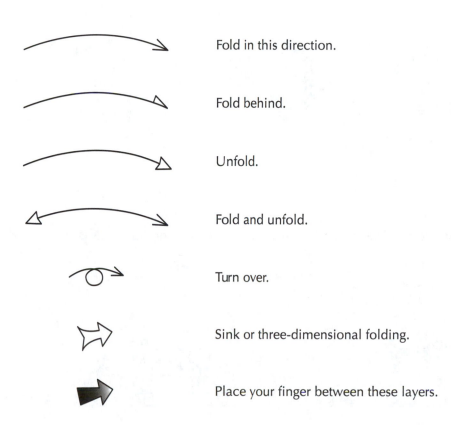

Fold in this direction.

Fold behind.

Unfold.

Fold and unfold.

Turn over.

Sink or three-dimensional folding.

Place your finger between these layers.

Basic Folds

Squash Fold

In a squash fold, some paper is opened and then made flat. The shaded arrow shows where to place your finger.

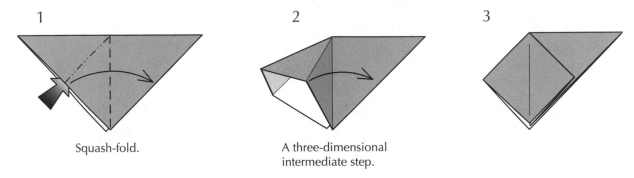

Squash-fold.

A three-dimensional intermediate step.

Inside Reverse Fold

In an inside reverse fold, some paper is folded between layers. Here are two examples.

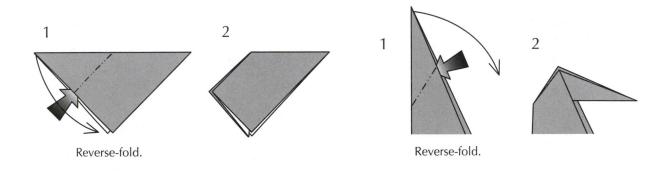

Reverse-fold.

Reverse-fold.

Sink Fold

In a sink fold, some of the paper without edges is folded inside. To do this fold, much of the model must be unfolded.

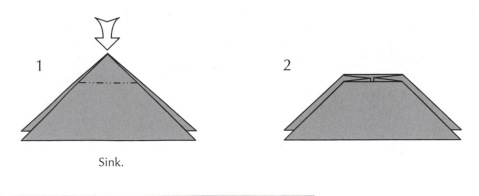

Sink.

Introduction

Polyhedra are beautiful shapes that have fascinated people throughout the ages. In ancient times, they were thought to possess magic powers. The Greeks knew of the five Platonic solids and associated them with the nature of the universe. Over time, more polyhedra were discovered, and they became better understood. Physical models of these geometric shapes have been formed in many media, including wood, metal, and paper. In this collection, you will learn to make many beautiful polyhedra, each by folding a single square sheet of paper. Also, you will see how origami skill, mathematics, and polyhedra can be combined in an elegant way.

I am dedicated to the pursuit of creating each origami model from a single uncut square. I developed methods for folding animals with that ideal at a time when using multiple sheets and cutting was prevalent. My techniques, discoveries, and ideals influenced many creators, leading to a more developed art. Now I wish to do the same with polyhedra.

There has been a movement, over the past 15 years or so, of creating geometric shapes with modular origami—the use of multiple sheets that are folded into identical units (modules) and then interlocked. For example, each edge of a polyhedron may be formed by one unit, and several units will interlock to form a face or vertex. I wanted to explore how these same shapes could be created from a single sheet of paper. Designing polyhedra from a single uncut square has led me to many new techniques and discoveries in origami that I wish to share with you.

My models are designed to be folded from standard origami paper (sometimes called kami). This paper comes in squares, typically 6 to 10 inches wide, that are white on one side and colored on the other. In the diagrams in this book, the colored side is represented by gray shading. Most of the models end up with only the colored side showing (and so single-colored paper could be used), but several designs create color patterns by exposing the white side as well as the colored side. Origami can also be folded from a variety of other papers, including standard notebook or printer paper, or from even thicker paper, using the technique of wet-folding.

This may be a good place to point out some folding techniques employed in this collection that may be less familiar to the casual folder. Most of the instructions begin with sequences of folds that are not folds in the final model but rather are used to find particular landmarks, such as one seventh of the edge of the paper or 30° through the center of the paper. Such sequences entail making small creases, such as folding the paper in half but only creasing on the left edge. In addition, I aimed to keep the faces of the final model with as few creases as possible, which led to some steps that require creasing only along part of a fold.

After the crease pattern defining the faces of the polyhedron is formed, the model becomes three-dimensional, typically at a step to either puff out or push in at some central point on the paper. The folding generally ends by tucking or locking one or more tabs. Throughout the folding instructions, symmetry plays a large role, which simplifies the folding.

The models range from simple to very complex, most being intermediate or complex. Still, none of the models are too difficult, and even the most complex has fewer than 60 steps. The models have been test-folded by many folders.

In Part I, "Designing Origami Polyhedra," there are references to several polyhedra whose directions are not given in this volume. My hope is to write a second volume of origami polyhedra that will include them.

The illustrations conform to the internationally accepted Randlett-Yoshizawa conventions. (See "Symbols" (page xi)). Origami paper can be found in many hobby shops or purchased by mail from OrigamiUSA, 15 West 77th Street, New York, NY 10024-5192.

Many people helped with this project. I thank Brian Webb and John Szinger for their continued support throughout this project. I thank Daniel Spaulding for his contributions. I thank John Szinger for folding some of the polyhedra featured on the cover. In particular, I thank my editors, Jan Polish, Charlotte Henderson, and Ellen Ulyanova. I also thank the many folders who proofread the diagrams.

John Montroll

www.johnmontroll.com

Designing Origami
Polyhedra

Polyhedra Overview

Polyhedra are three-dimensional shapes composed of polygons, two-dimensional shapes composed of straight lines. Here we present some of the names associated with polyhedra.

Convex Polyhedra

Take any two points on the surface of a polyhedron and picture a line connecting them (as in the first picture below). A polyhedron is convex if all such lines are either on the surface of the polyhedron or are completely inside the polyhedron.

Icosahedron

Pentagonal
Pyramid

Hexagonal
Dipyramid

Concave Polyhedra

If there are any lines connecting two points on the surface that are outside the polyhedron (as in the first picture below), even partially, then the polyhedron is concave, or nonconvex.

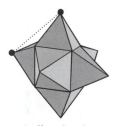

Stellated Cube
(Star)

Cubehemioctahedron
(Dimpled)

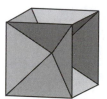

Sunken Cube
(Sunk)

Platonic Solids

A Platonic solid is a convex polyhedron composed of identical regular polygons with identical vertices. Thus, all of the edges are the same length, the same number of edges meet at each vertex, and all the faces are the same shape. There are five Platonic solids: the tetrahedron, cube, octahedron, icosahedron, and dodecahedron.

Tetrahedron

Cube

Octahedron

Icosahedron

Dodecahedron

Archimedean Solids

The Archimedean polyhedra are convex solids that are made from two or more types of regular polygons and that have identical vertices. (This is similar to the definition of Platonic solids except all the faces are *not* the same shape.) Here are some of the 13 Archimedean solids.

(A *truncated* polyhedron is created by slicing off the corners, leaving regular polygons in place of the original vertices.)

Truncated Tetrahedron

Cuboctahedron

Snub Cube

Rhombicuboctahedron

Truncated Octahedron

Icosidodecahedron

Truncated Icosahedron

Prisms

A prism is a polyhedron where the cross sections are parallel to its base. It is composed of two *n*-gons and *n* identical rectangles.

Triangular Prism

Cube

Pentagonal Prism

Hexagonal Prism

Antiprisms

The two bases of a prism line up evenly with each other. For an antiprism, the two bases are offset, so that it is composed of two identical *n*-gon bases and 2*n* lateral triangles going around its equator.

Tall Triangular
Antiprism

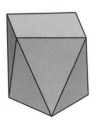
Tall Square
Antiprism

Golden Pentagonal
Antiprism

Tall Hexagonal
Antiprism

Pyramids

A pyramid has a polygon base and triangles whose vertices meet at the top, somewhere above the center of the base.

Triangular Pyramid

Egyptian Pyramid

Hexagonal Pyramid

Duals

The dual of a polyhedron is a polyhedron where the faces and vertices are switched. For many somewhat round-shaped polyhedra, the dual can be obtained by placing a point in the center of each face, connecting the points to form a new polyhedron, and scaling it so both polyhedron and dual are inscribed in the same sphere.

The cube and octahedron are dual pairs.

The icosahedron and dodecahedron are dual pairs.

The tetrahedron is its own dual.

In general, the process of constructing the dual is called polar reciprocation. For nonrounded shapes, the dual is found by drawing lines from the center of the polyhedron through the centers of each face. The closer the face is to the center of the solid, the further the vertex will be in the dual.

For example, consider the uniform pentagonal prism. Two sides are pentagons and five are squares. The dual of this somewhat squat shape is the pentagonal dipyramid, a longer shape.

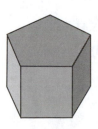

The uniform pentagonal prism and pentagonal dipyramid are dual pairs.

The square antiprism and square trapezohedron are dual pairs.

Catalan Solids

The Catalan polyhedra are the duals of the Archimedean solids. Each one is convex with identical faces, though none of them are regular polygons. The vertices are not identical. Here are some of the 13 Catalan solids.

Triakis Tetrahedron

Rhombic Dodecahedron

Triakis Octahedron

Triakis Cube

Dipyramids

Dipyramids, or diamonds, are the duals of prisms. Dipyramids are composed of a pair of identical pyramids joined at the base. All the sides are identical isosceles triangles.

Squat Silver
Square Dipyramid

Hexagonal
Dipyramid

Heptagonal
Dipyramid

Decagonal
Dipyramid

Trapezohedra

Trapezohedra, or antidiamonds, are the duals of antiprisms. The sides are identical kite-shaped quadrilaterals where three of the angles are the same.

Triangular Trapezohedron
(Cube)

Square
Trapezohedron

Pentagonal
Trapezohedron

Octagonal
Trapezohedron

Design Factors and Techniques

Designing origami polyhedra is an interesting challenge. It combines origami skill with mathematics. There are many factors to take into consideration, and it is difficult to satisfy all of these criteria at once:

- The faces should have no lines or creases, or as few as possible.

- The paper should be used efficiently to yield the largest possible result.

- The model should lock cleanly and retain its shape.

- The folding structure should maximize symmetry.

- The folding sequence should be as simple as possible and accessible to the general origami public.

Here is an outline of the design process.

First, work out several two-dimensional layouts of the shape. Taking symmetry into account, (from best to worst) use square symmetry, even/odd, odd, even, radial, or none. In general, each face of the polyhedra should have no lines. But for many solids, this could be too restricting, so think of layouts where some sides are split (see my example of one of the cube layouts where one side has an X pattern). This could simplify symmetries and the folding. Using a computer graphics application, it is easy to draw many layouts. Draw small pictures of each of the faces, and arrange them in different ways, saving all the arrangements.

Second, configure the layout on a square to form the crease pattern. For some, the vertices could meet the edge of the paper. Occasionally an edge from the layout will meet the edge of the paper. With a graphics application, simply draw several possibilities. For a given layout, try different rotations and sizings to form the crease pattern. From these, make the crease pattern as large as possible and print it. Just cut out the square and fold using the lines. If it works, continue to the next phase.

Third, find the landmarks necessary to generate the creases. Sometimes the landmarks are easy to find. Often some math is used and numbers are calculated. For some of the numbers, it is easy to find the landmarks. But for many, Robert Lang's ReferenceFinder software is used to generate the folding procedure.

Now comes the rest of the folding procedure, including lock. Generally, models with sunken sides hold easier than convex polyhedra, allowing for efficiency.

Once you have worked one out, you are ready for another. Some models can be used as bases to create more polyhedra. For example, a cube can be folded into a cubehemioctahedron by sinking the vertices.

Here we explore in more depth the general design concepts outlined above. Later, in "Math and Design" (page 19), we will dive into details of how the designs are realized, as outlined in the third step.

Work out several 2D layouts.

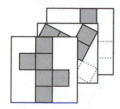

Place the layouts on square paper.

Find the landmarks.

Complete the folding.

Create related models.

Layout

The folding pattern for a polyhedron begins with an "outline": a two-dimensional layout of all of its faces, similar to what you would use if you were going to create the polyhedron by cutting out the layout and taping certain edges to complete the model. One could find such a layout online, place it on a square, and fold it. Most likely, however, it would lead to a rather clumsy model. By understanding the structure better, a more elegant design can be realized.

Let's define a *layout* to be the orientation of the faces of a polyhedron in two dimensions. A *crease pattern* is the layout on a square to be folded. For elegant designs, it is best not to split the faces. Also, we will need to leave some of the paper for the tabs (see "Tabs" (page 11) for more details).

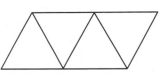

The layout of a tetrahedron shows a band of four triangles.

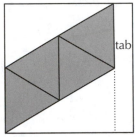

The crease pattern shows the position of the layout on a square.

Tetrahedron

Symmetry

Symmetry is very important to the layout. Through symmetry, the model is easier to design and fold. Here are different forms of symmetry for the crease pattern of polyhedra. They will be explained in detail.

No Symmetry

When the crease pattern is rotated or flipped and cannot be the same as its original pattern, then there is no symmetry. This form is the least useful. One of the few models I have designed with no symmetry is the Tetrahedron shown above.

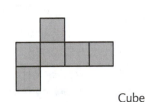

Cube

Radial Symmetry

A pattern has radial symmetry if it is the same when rotated by some angle (divisible by 360°) other than 90° or 180°. It is useful when the paper is a polygon other than a square. However, when folding from a square, this method is generally not ideal. It could be useful for folding a square into a polygon and then into a simple polyhedron such as a pyramid.

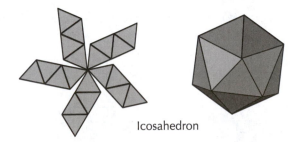

Icosahedron

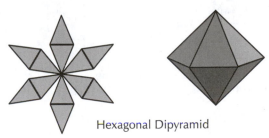

Hexagonal Dipyramid

Even Symmetry

The reflection along a line through the middle of the pattern is the mirror image when that pattern has even symmetry. This method is generally not very useful. The main use for even symmetry is for polyhedra whose sides are different shapes. Some of the pyramids and prisms I have designed use even symmetry.

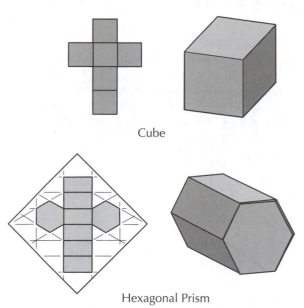

Cube

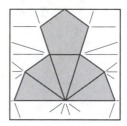 

Golden Pentagonal Pyramid

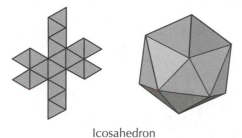

Hexagonal Prism

Odd Symmetry

For odd symmetry, the pattern is the same when rotated 180°.

This is the single most important form of symmetry used for polyhedra. As I design models, I generally only consider odd symmetry. There are some polyhedra, such as the pyramid above, that cannot use odd symmetry.

Most animals follow even symmetry, whereas odd symmetry is preferred for polyhedra.

When using odd symmetry, there is a line going through the center at some angle. This is generally one of the first creases to be folded.

The number of identical faces of the polyhedra has to be an even number.

All futher symmetries to be mentioned contain odd symmetry.

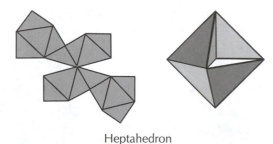

Icosahedron

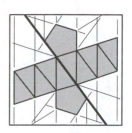

Heptahedron

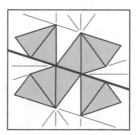 

The bold line goes through the center for the Golden Pentagonal Dipyramid.

Golden Pentagonal Antiprism

Even/Odd Symmetry

As the name suggests, with this symmetry the crease pattern is both even and odd. When even/odd symmetry applies, it can be very useful. The number of identical faces of the polyhedron has to be a multiple of four.

Generally, this type of pattern really has just odd symmetry: when it comes to including the tabs, the symmetry becomes odd.

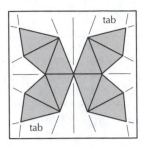

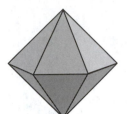

The tabs make an otherwise even/odd Hexagonal Dipyramid have odd symmetry.

Hexagonal Dipyramid in a Sphere

Square Symmetry

For square symmetry, the pattern is the same when rotated 90°. This is a special case of having both radial and odd symmetry.

When possible, this is often the best choice for a folding pattern from a square piece of paper. It has the potential for producing the most efficient layout. Because of repetition, only 1/4 of the folding needs to be diagrammed. Typically, there is a vertex at the center of the paper.

The number of identical faces of the polyhedra has to be a multiple of four.

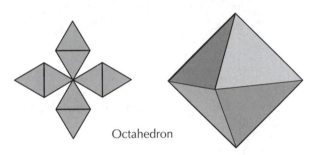

Octahedron

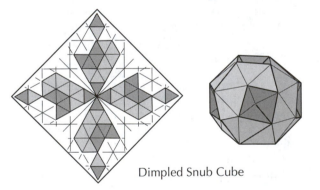

Dimpled Snub Cube

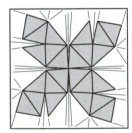

Triakis Cube

3/4 Square Symmetry

This unusual pattern at first appears to have no symmetry. But when you consider the 1/4 of the paper that is hidden in the final model, the crease pattern has square symmetry.

It is useful for the cube and related polyhedra. The number of identical faces of the polyhedra has to be a multiple of three.

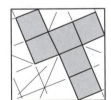
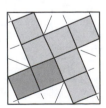

The layout for the Cube shows 3/4 square symmetry. The middle drawing shows the underlying square symmetry.

Tabs

The tabs and locking methods are among the trickiest of the problems that arise when designing origami polyhedra. Everything hinges around them. A large, seemingly efficient design might not have enough tab to lock the faces in place. An efficient use would have just the right amount of tab.

The tabs are dependent on the symmetry used. Odd symmetry often allows for two tabs to make a knot. This is used for several of the dipyramids.

Square or 3/4 square symmetry leads to the twist lock, a very efficient method. Three or four sides meet at a vertex. The tabs spiral inside to lock the model.

Generally, the tabs are found on the corners of the square. Their size is arbitrary. In many designs, I work out a layout to be as large as possible. If the model does not hold well, then I work out a crease pattern with the same layout but either smaller or at a different angle to allow more tab.

Consider the octahedron. To make the largest model, the eight triangles are arranged using square symmetry so that four vertices meet the corners of the square. However, the crease pattern provides no tab. By rotating the layout a little, each corner has some excess to form a tab. To finish the folding, the four tabs meet and close with a twist lock. This efficient lock allows for a large octahedron.

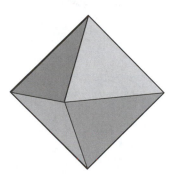

Octahedron

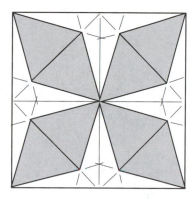

This crease pattern with square symmetry would make the largest octahedron, but there is no tab.

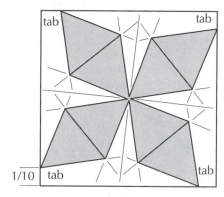

By rotating the layout, the four corners become tabs. This leads to the twist lock.

The folding method for the Pentagonal Dipyramid shows a typical lock from a layout using odd symmetry.

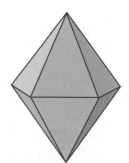

Pentagonal Dipyramid

This is one of the last steps in the folding. The model is still 2D. The tabs are shown on both sides. The model now opens and becomes 3D.

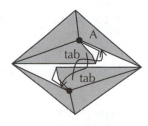

Tuck and interlock the tabs. The dots will meet. This type of lock forms a knot.

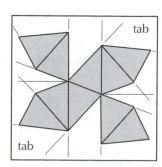

The crease pattern shows odd symmetry with the two tabs.

Optimizing the Crease Pattern

Optimizing refers to finding the largest model with a good tab while keeping the folding as simple as possible. Symmetry plays an important part.

Given two crease patterns for the same polyhedron, the efficiency in terms of size can be found. Here are three ways of measuring that efficiency:

1. Measure the area of the faces of the model and divide by the area of the square paper. The larger the number, the larger the model.

2. Calculate the length of a side of one of the faces in a 1 × 1 square. Compare the lengths from two crease patterns. The larger length yields a larger model.

3. Represent both crease patterns using faces of the same size. The pattern embedded in a smaller square yields a larger model.

Compare the following two crease patterns for the octahedron. One uses odd symmetry, the other square symmetry. For a 1 × 1 square, the length s of a side of the faces is given. Also, the triangular faces are all the same size.

Octahedron

$$s = 1/\left(2\sqrt{3}\right) \approx .2887$$

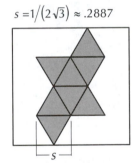

Odd symmetry.

$$s = \sqrt{123}/30 \approx .3697$$

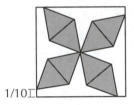

1/10

Square symmetry. This crease pattern yields a larger octahedron

Let's look at two examples, the Cube and the Icosahedron, to better explore what it means to optimize a crease pattern.

Optimizing the Cube

Let's optimize to find the crease pattern that generates the largest cube. The six faces suggest using even, odd, or 3/4 square symmetry. For each pattern, the length of the side will be given.

The sides have length .25 in these crease patterns.

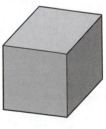

Cube

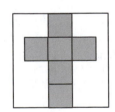

Cross layout, even symmetry.

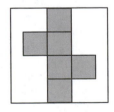

Odd symmetry.

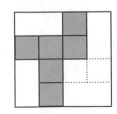

3/4 square symmetry.

By rotating the layout with 3/4 square symmetry by 22.5°, the sides increase from .25 to .2706. This model is diagrammed in "Cube Design" (page 83). It closes with a three-way twist lock.

$$s = 1/(4\cos(22.5°)) \approx .2706$$

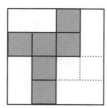 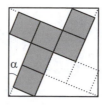

$$\alpha = 22.5°$$

In the rotated version, three vertices meet the edges of the square. The layout can be rotated a bit more so that six vertices meet the edges. This would optimize the 3/4 symmetric layout with respect to size. The cube it generates would be larger, but there would not be enough tab to hold it together. Thus, this crease pattern is not used for the cube. However, it is used to optimize other related polyhedra. This includes the Triakis Tetrahedron and Cubehemioctahedron.

$$s = 1/(4\cos(\arctan(.5))) \approx .2795$$

Triakis Tetrahedron   Cubehemioctahedron

Now consider placing a band of four square faces along the diagonal with even or odd symmetry. This increases the sides of each square. Unfortunately, there is not enough tab.

$$s = \sqrt{2}/5 \approx .283$$

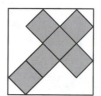 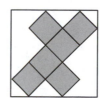

Even symmetry.   Odd symmetry.

Though we are used to placing sides next to each other in a layout, they do not have to be. By rotating two square faces so they only meet at a vertex, we can make another layout with the same efficiency. This crease pattern yields a cube that can lock. Thus, this is the most efficient cube. It will be diagrammed in "Prism Design " in a future volume on origami polyhedra.

$$s = \sqrt{2}/5 \approx .283$$

Even/odd symmetry optimizes the cube.

Optimizing the Icosahedron

With its 20 sides and somewhat ball-like shape, this model is a challenge to design. Before taking symmetry into account, let us try a few ideas.

Icosahedron

It would be very convenient to make a layout with one face in the center and the rest revolving around it. Here is a possible layout.

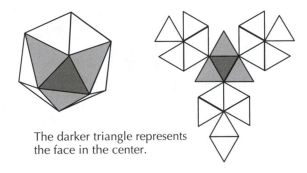

The darker triangle represents the face in the center.

The layout has three-way radial symmetry plus one extra face on the bottom. In general, placing a face in the center does not optimize the layout. The few times where it does is when there are an odd number of faces or the model does not have odd symmetry.

Consider the standard layout given in books.

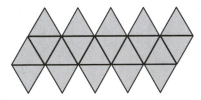

Standard layout.

The difficulty with this is it does not fill out the square paper very well and the locking would be difficult. Let's try radial symmetry.

Radial symmetry.

It would again be difficult to place this layout in a square. Also, the locking would involve five vertices to meet, which does not hold together well.

Let's try the band method, as with the cube. The equator has a band of ten triangles that will be used in the layout. Place the band along the diagonal in the crease pattern. There will be five triangles on either side. This has odd symmetry and fills the square well. It also has plenty of space for tabs, allowing the model to lock well. Because of the 15° lines starting from the origin, it is easy to set up the crease pattern and fold. (We will discuss folding angles in more detail in "Math and Design" (page19).)

The equator is shaded.

Layout with a band of ten triangles.

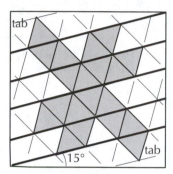

The 15° lines are bolded in the crease pattern.

The layout could be made larger in the square paper, but the landmarks—key elements upon which the rest of the folds are based, discussed in "Math and Design" (page 19)—would be difficult to find. Also, finding landmarks could lead to extra creases that show on the faces, reducing the crisp look of the final model. Also, with a larger layout, there would be less tab. Thus, this is my optimal crease pattern for the Icosahedron.

Groups of Polyhedra

We have seen that the band method optimizes the crease pattern for the cube and the icosahedron. Does the band method also work for other polyhedra? Can we sort polyhedra into groups so that each group has a similar crease pattern? If so, then once we determine an efficient pattern for one polyhedron in that group, it may lead us easily to efficient patterns for the others. By understanding the structure of the two-dimensional layout with respect to symmetry, tabs, and optimization, it is possible to develop methods of making related polyhedra.

Dipyramids

The dipyramids have a unified structure. I came up with a formula relating the angles of the triangular faces to the proportions of the finished model. This made it easy to design a large collection of dipyramids.

Let H = height divided by diameter, α = angle at top of each triangle, n = number of sides of the polygonal base.

$$H = \tan\left(\arccos\left(\frac{\sin(\alpha/2)}{\sin(180°/n)}\right)\right)$$

and

$$\alpha = 2\arcsin[\sin(180°/n)\cos(\arctan(H))]$$

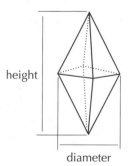

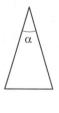

This led to two general crease patterns depending on whether the polygonal base has an even or odd number of sides.

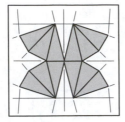

The Hexagonal Dipyramid has even/odd symmetry.

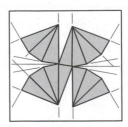

The Heptagonal Dipyramid has odd symmetry.

Antiprisms

For antiprisms, I came up with a general structure. Using this structure, I easily designed several of them.

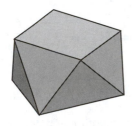

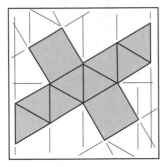
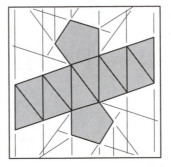
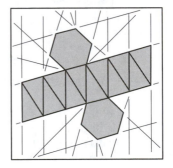

Square Antiprism with crease pattern.

Golden Pentagonal Antiprism with crease pattern.

Hexagonal Antiprism with crease pattern.

The method for developing antiprism crease patterns goes as follows:

Leave some space on the left and right sides of the square for the tabs. Fold vertical lines evenly spaced and make a band of triangles, which make up the equator of the antiprism, going through the center of the square. Add the two remaining polygon faces, one on either side of the band. All of this is done using odd symmetry.

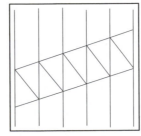
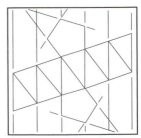

Vertical lines with some space on the left and right for tabs.

Band of triangles.

Two polygons are added.

Tetrahedra with Color Patterns

The Tetrahedron has a band of four triangles in its crease pattern. The same band can be used to create tetrahedra with color patterns.

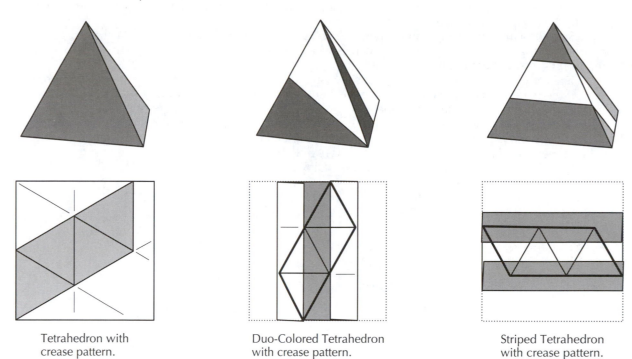

Tetrahedron with crease pattern.

Duo-Colored Tetrahedron with crease pattern.

Striped Tetrahedron with crease pattern.

Related to the Cube

This group of polyhedra has the same surface as the cube. Thus, by folding a cube and sinking vertices or making other changes, a variety of polyhedra can be created.

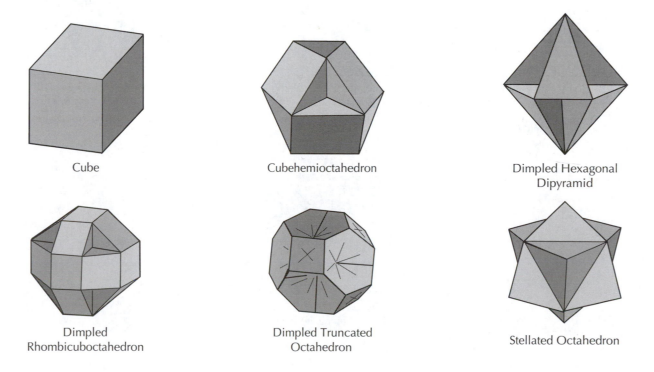

Cube

Cubehemioctahedron

Dimpled Hexagonal Dipyramid

Dimpled Rhombicuboctahedron

Dimpled Truncated Octahedron

Stellated Octahedron

Folding from a Crease Pattern

Suppose you have a crease pattern and you want to fold the polyhedron from it. It will take some experimenting to find the best way to do so. Here are a few ideas.

Start with the white side of the paper. The crease pattern shows the orientation of the faces. Valley-fold along the edges of the faces. There are places where faces meet at a vertex with some angle between them. Bisect those angles with mountain folds.

Basically, the part with the faces will show, and the rest will be hidden inside the polyhedron.

If there are lines that go through the exact center of the square, fold those as part of the first folds. They will be valley or mountain folds depending on whether they are on the edges of the faces.

As the model is being folded, try to avoid creasing over the faces. See if there is some excess paper from opposite corners that can be used as tabs. Sometimes, I draw the edges of the faces so that when I fold the model I will know what to show and what to hide.

Here is the crease pattern for a Pentagonal Dipyramid.

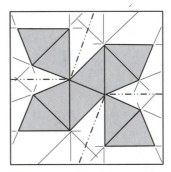

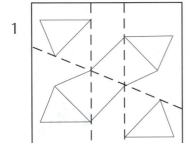

1. Begin by valley-folding along the crease going through the center. Then valley-fold along the two vertical lines.

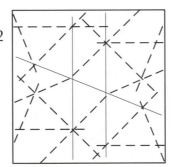

2. Continue valley-folding along the edges. The lines can extend beyond the vertices as long as they do not pass through the faces.

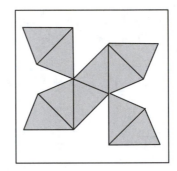

3. Bisect the angles between the faces with mountain folds.

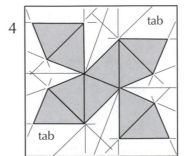

4. Collapse along the creases. There is some paper on opposite corners that can form tabs.

Math and Design

Once we have determined how to best arrange a layout on our square paper, we have to figure out how to perform the folds needed to generate the crease pattern. We will need to fold particular angles or find a particular length along a side of the square. To do this, we employ mathematics. A square sheet of paper lends itself well to math, especially when designing or understanding origami polyhedra. In general, algebra, geometry, and trigonometry are used.

Typical uses of math in designing polyhedra are the following:

- to find the lengths of lines and the values of angles,

- to find the landmarks of a crease pattern,

- to divide the paper into $n$ths (such as thirds).

We will first outline some basics of mathematics on the square.

Basics

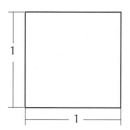

Consider a square to be 1 unit by 1 unit in length. The length of the diagonal is found by the Pythagorean Theorem, using the right triangle formed by two adjacent edges and the diagonal as the hypotenuse:

$$a^2 + b^2 = c^2$$
$$1^2 + 1^2 = c^2$$
$$c = \sqrt{2} \approx 1.4142136$$

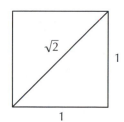

Any point on the square is defined by (x, y) where both x and y are values from 0 to 1. In the third image on the right are some examples.

Any crease defines a line. Lines can be expressed by

$$y = mx + b,$$

where m = slope, or the change in y divided by the change in x, and b = y-intercept, or where the line meets the left side of the paper.

Equivalently, a line can be expressed as

$$y - y_1 = m(x - x_1),$$

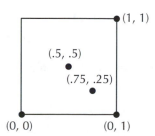

where (x_1, y_1) is a point on the line.

The line between points $(0, .25)$ and $(1, 1)$ is

$$y = .75x + .25, \quad \text{since}$$
$$m = \frac{y_2 - y_1}{x_2 - x_1} = \frac{1 - .25}{1 - 0} = .75$$

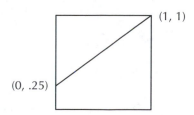

When folding practically anything, each crease can be defined by an equation of a line. Fortunately, the process of folding is oblivious to the math.

Landmarks

When designing origami models, there is often a landmark that is the key to folding the model. Given a crease pattern, the landmark can be found mathematically. The landmark can be represented as a point (x, y) on a 1×1 square. Quite often, the landmark occurs on the edge of the square.

For many models, the landmarks are divisions of the paper into *n*ths, such as thirds. Other landmarks deal with angles, especially going through the center of the paper.

Examples:

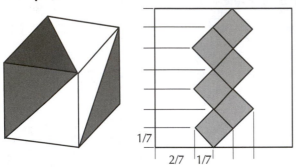

The crease pattern for Triangles on Cube shows divisions of 1/7.

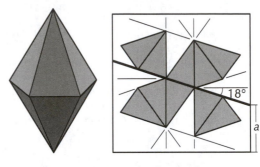

In the crease pattern for the Golden Pentagonal Dipyramid, there is a line through the center at 18° from the horizontal line. The landmark *a* is calculated to be .33754.

Other landmarks require more calculations.

Example:

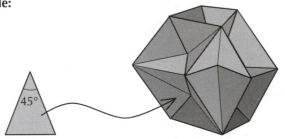 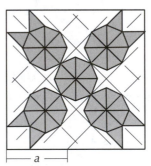

The Sunken Rhombic Dodecahedron is composed of 48 triangles. Each face is an isosceles triangle with an apex angle of 45°. Once the landmark *a* is found, the model can be folded.

Calculation of Landmark Using Algebra

Algebra is often used to find landmarks.

To fold a Triakis Tetrahedron or Cubehemioctahedron, the following layout is used. To make the most efficient use on a square, it is rotated so the outermost vertices meet the edges of the square sheet.

Triakis Tetrahedron

Cubehemioctahedron

Layout

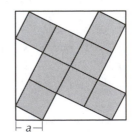

Layout rotated for best placement on square.

The problem is to find landmark *a*. Once known, the model can be folded. After some algebra, it is determined that $a = .25$. That landmark is easy to fold. Next is the math used and the first few diagrams.

Find length a given a 1×1 square.

Let s = length of the side of the small squares.

$$c^2 = s^2 + (2s)^2 = 5s^2$$
$$c = \sqrt{5}s$$

Triangles T_1 and T_2 are similar, so

$$\frac{a}{b} = \frac{2s}{s}$$
$$a = 2b$$
$$b = \frac{a}{2} \text{ and}$$
$$\frac{a}{s} = \frac{2s}{c}$$
$$ac = 2s^2$$
$$a = \frac{2s^2}{c} = \frac{2s^2}{\sqrt{5}s} = \frac{2s}{\sqrt{5}}.$$

Since $b = a/2$, then $b = s/\sqrt{5}$.

For a 1×1 square,

$$a + c + b = 1$$
$$\frac{2s}{\sqrt{5}} + \sqrt{5}s + \frac{s}{\sqrt{5}} = 1$$
$$\frac{8}{\sqrt{5}}s = 1$$
$$s = \frac{\sqrt{5}}{8}. \text{ So,}$$
$$a = \frac{2s}{\sqrt{5}} = \frac{2}{\sqrt{5}}\frac{\sqrt{5}}{8} = \frac{1}{4}$$
$$a = .25.$$

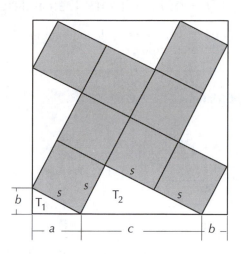

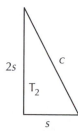

Triangles T_1 and T_2 are similar.

To see how this landmark is used, let's look at the first six steps of the Triakis Tetrahedron. Note the .25 marks in Step 1. In Step 6, we can make out the outline of the underlying layout.

Triakis Tetrahedron

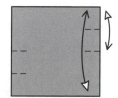

1

Make small marks by folding and unfolding in quarters.

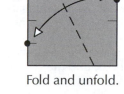

2

Fold and unfold.

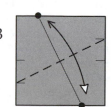

3

Fold and unfold.

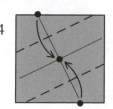

4

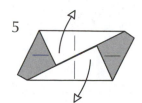

5

Unfold.

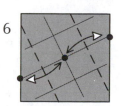

6

Fold and unfold.

Calculation of Landmark Using Trigonometry

Trigonometry is also used to calculate landmarks. Here are some basic trigonometric formulas that we use often in such calculations.

Right triangle.

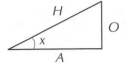

Adjacent
Opposite
Hypotenuse

To find length of a line:

$$A^2 + O^2 = H^2$$

$$\sin(x) = O/H$$
$$\cos(x) = A/H$$
$$\tan(x) = O/A$$

To find angle:

$$\arcsin(O/H) = x$$
$$\arccos(A/H) = x$$
$$\arctan(O/A) = x$$

Any triangle.

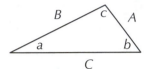

Law of Sines

$$\frac{\sin(a)}{A} = \frac{\sin(b)}{B} = \frac{\sin(c)}{C}$$

or

$$\frac{A}{\sin(a)} = \frac{B}{\sin(b)} = \frac{C}{\sin(c)}$$

Example:

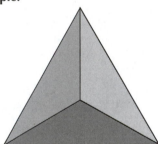

Triangular Dipyramid

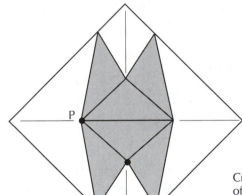

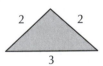

Crease pattern and one of the triangular faces.

This Triangular Dipyramid is composed of six triangles. Each triangle has sides proportional to 2, 2, and 3. In the crease pattern there are three landmarks, shown by the dots. Knowing landmark P is sufficient for finding the rest and folding the model.

Given the crease pattern and the proportions of each triangle to be 2, 2, and 3, find the coordinates of landmark P in a 1 × 1 square.

1. Find the angles of the triangular faces.

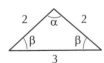

$$\sin(\alpha/2) = 1.5/2 = .75$$
$$\alpha/2 = \arcsin(.75)$$
$$\alpha \approx 97.18°$$

$$\alpha + 2\beta = 180°$$
$$\beta = 41.41°$$

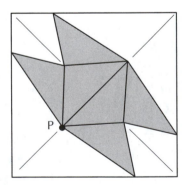

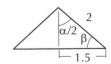

2. Find lengths a and c. We get some information from the diagonal:

$$a + 2c = \sqrt{2}$$

We get more information by looking at triangle T.

- Find angle λ.

$$2\beta + \lambda = 180°$$
$$\lambda = \alpha \approx 97.18°$$

- Find angle ϕ.

$$\phi = 180° - 45° - \lambda$$
$$\phi \approx 37.82°$$

- Use Law of Sines to find c.

$$c/\sin(\phi) = a/\sin(45°)$$
$$c = a\sin(\phi)/\sin(45°)$$
$$\approx .8671567a$$

Now we use the diagonal equation to find length a.

$$a + 2c = \sqrt{2}$$
$$a + 2(.8671567)a = \sqrt{2}$$
$$a = \sqrt{2}/(1 + (2)(.8671567))$$
$$= .5172097$$

Then,

$$c = .8671567a$$
$$= .4485018$$

3. Find the coordinates of P.

Since P is along the diagonal, $x = y$ and c is the hypotenuse of an isosceles triangle:

$$c^2 = x^2 + y^2 = x^2 + x^2 = 2x^2$$
$$x = c/\sqrt{2}$$

so, P $= (c/\sqrt{2}, c/\sqrt{2}) \approx (.3171387, .3171387)$

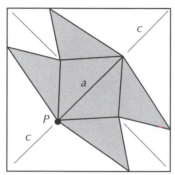

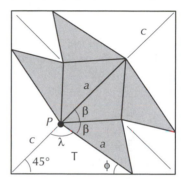

A folding sequence locates the landmark (.3171387, .3171387). Here are the first few steps of the Triangular Dipyramid.

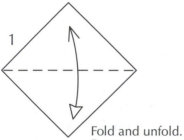

1

Fold and unfold.
Rotate.

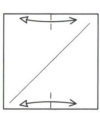

2

Fold and unfold
at the edges.

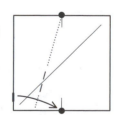

3

Bring the left edge to the lower dot. Crease at the intersection.

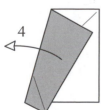

4

Unfold and
rotate 180°.

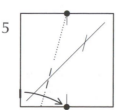

5

Repeat steps 2–4.
Turn over and rotate.

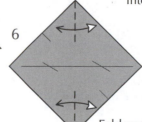

6

Fold and unfold but do
not crease in the center.

Robert Lang's ReferenceFinder

Once the landmark is found, then a folding method is used to locate it. As an origami designer, I want my models to be realized by folding alone without a ruler or protractor, and even with no awareness to the folder that there might have been pages of calculations. Sometimes, the landmarks are easy enough to find with a few folds. But quite often, it would be difficult to think of an efficient folding procedure. So I use Robert Lang's ReferenceFinder computer program, http://www.langorigami.com/referencefinder.htm.

ReferenceFinder gives an efficient folding sequence to locate any point on a square with a small error. The folding sequence typically takes only three to five steps.

When designing the Golden Pentagonal Antiprism, the following layout was used.

Golden Pentagonal
Antiprism

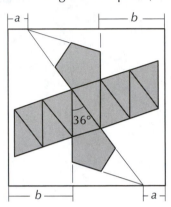

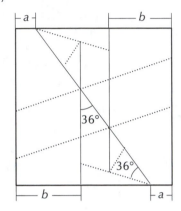

Landmarks *a* and *b* are all that is needed to set up the folds.

Length *a* defines the line going through the center of the paper at 36° from a vertical line. It was easy enough to calculate.

$$a = .5(1 - \tan(36°)) \approx .1367287$$

ReferenceFinder found a simple folding sequence to locate landmark *a*.

Length *b* defines an isosceles (36°, 36°, 108°) triangle and pentagon below and above the center band. It was difficult to calculate. Yet it was determined that

$$b = \sqrt{2} - 1 \approx .41421356$$

which is an easy landmark to find.

According to ReferenceFinder, landmark *a* can be found this way:

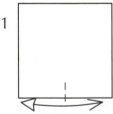

Fold in half at the bottom.

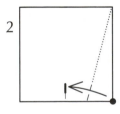

Unfold.

$a \approx .1367287$

Fold and unfold.

Landmark *b*, though difficult to calculate, is easy to find and did not require ReferenceFinder.

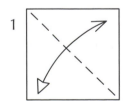

Fold and unfold
along the diagonal.

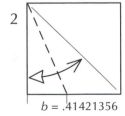

$b = .41421356$

Fold and unfold
to the diagonal.

Dividing a Square into *n*ths

First, some specific cases will be shown. Then some general methods will be given.

Divisions of 1/3

1	2	3	4	5
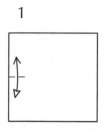	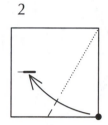	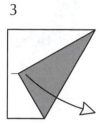	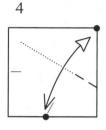	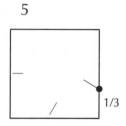
Fold and unfold in half on the left.	Crease at the bottom.	Unfold.	Fold and unfold on the right.	The 1/3 mark.

Divisions of 1/5

Method 1: This is an approximation with a division of 0.198912. This is quick and simple to fold and generally has good enough precision (99.5%).

1	2	3	4	5
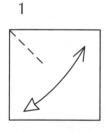	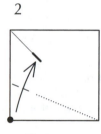	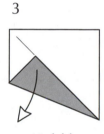	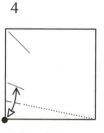	
Fold and unfold creasing along part of the diagonal.	Crease on the left.	Unfold.	Fold to the crease and unfold.	The 1/5 mark.

Method 2: This is exact but slightly more difficult to fold.

1	2	3	4	5
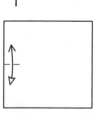		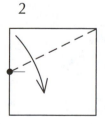	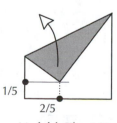	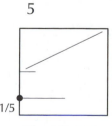
Fold and unfold in half on the left.		Fold and unfold on the right.	Unfold. The 1/5 and 2/5 locations are found.	The 1/5 mark.

Divisions of 1/6

The first three steps are the same as those used to find the 1/3 mark.

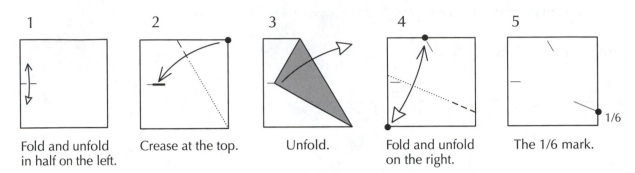

1
Fold and unfold in half on the left.

2
Crease at the top.

3
Unfold.

4
Fold and unfold on the right.

5
The 1/6 mark.

Divisions of 1/7

This is an example of the diagonal method for dividing the square into *n*ths as shown on the next page.

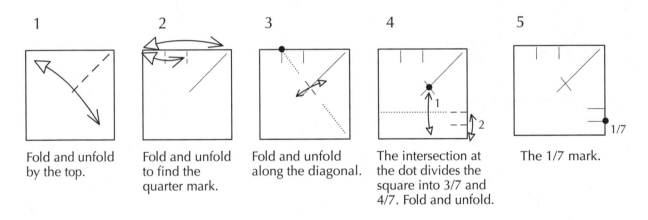

1
Fold and unfold by the top.

2
Fold and unfold to find the quarter mark.

3
Fold and unfold along the diagonal.

4
The intersection at the dot divides the square into 3/7 and 4/7. Fold and unfold.

5
The 1/7 mark.

Divisions of 1/8

The standard method for finding the 1/8 mark is to divide the paper in half three times. That would take three folds. This clever method requires only two folds.

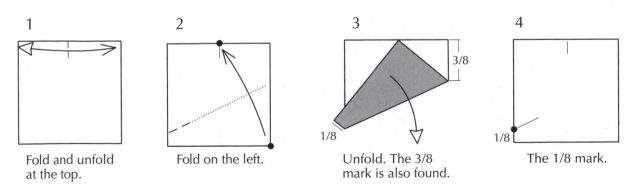

1
Fold and unfold at the top.

2
Fold on the left.

3
Unfold. The 3/8 mark is also found.

4
The 1/8 mark.

Divisions of 1/9

This is an example of the edge method for dividing the square into n^2 parts as shown on page 30.

1

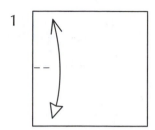

Fold and unfold on the left.

2

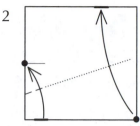

Bring the lower right corner to the top edge and the bottom edge to the left center. Crease on the left.

3

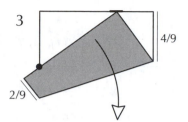

Unfold. The 2/9 and 4/9 marks are found.

4

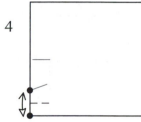

Fold and unfold.

5

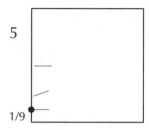

The 1/9 mark.

Diagonal Method for Dividing the Square into *n*ths

In this method, two lines are folded, and their intersection divides the paper into a desired fraction. One of the lines is along the diagonal of the square.

To use this method, we need to know the powers of 2. These are the numbers $2^1 = 2$, $2^2 = 4$, $2^3 = 8$, $2^4 = 16$, $2^5 = 32$, and so on. (Note that we can always divide a square into a fraction of a power of 2 by successively folding in half: see, for instance, Step 1 in the example.)

Example: Divide into 11ths.

We begin with two numbers that add up to 11. One of the numbers is the greatest power of 2 less than 11, which is 8. The other is 3 = 11 − 8.

Make a fraction of the two numbers. This would be 3/8. Find height 3/8 on the right edge, and fold a crease from the lower left corner to the 3/8 mark. Fold a downward diagonal crease. These two creases intersect at (8/11, 3/11).

1

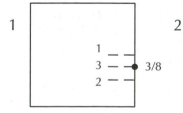

Fold and unfold in half three times to find the 3/8 mark.

2

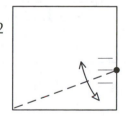

Fold and unfold.

3

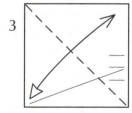

Fold and unfold.

4

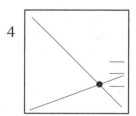

The dot is at (8/11, 3/11).

Let us check that intersection point. Our two creases are the lines $y = (3/8)x$ and $y = -x + 1$. They intersect at

$$-x + 1 = (3/8)x$$
$$1 = (11/8)x$$
$$= 8/11$$

and $y = (3/8)(8/11) = 3/11$.

Diagonal method for dividing the square into *n*ths.

Given n, find a and b so $n = a + b$ where $a = 2^m$ is the largest power of $2 < n$.

Example: $n = 11$ then $a = 8$, $b = 3$.
$n = 25$ then $a = 16$, $b = 9$.

1 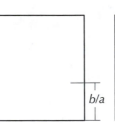 2 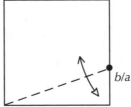 3 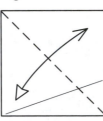 4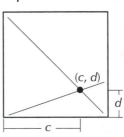

Find the location b/a on the right edge.

Fold and unfold from the lower left corner to the right edge at b/a.

Fold and unfold along the diagonal.

The intersection is at $(a/(a + b), b/(a + b))$.

Example: Divide into 27ths.

$$n = a + b$$
$$27 = 16 + 11$$
$$b/a = 11/16$$

1 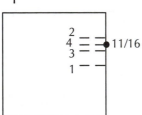 2 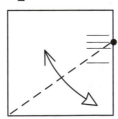 3 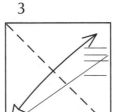 4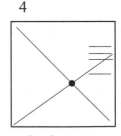

Fold and unfold in half four times to find the 11/16 mark.

Fold and unfold.

Fold and unfold.

The dot is at $(16/27, 11/27)$.

This method does not work well if b/a is small because of the small angle, which is physically difficult to fold.

Derivation of the diagonal method for dividing the square into *n*ths.

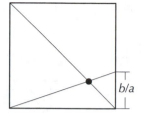

Find the intersection of the two lines:
Given a 1 × 1 square, the equation of the downward diagonal is

$$y = -x + 1$$

The equation of the line from the origin to (1, *b/a*) is

$$y = \frac{b}{a}x$$

Solve for *x* and *y*:

$$y = -x + 1,$$

$$y = \frac{b}{a}x,\text{ so}$$

$$\frac{b}{a}x = -x + 1$$

$$\frac{b}{a}x + x = 1$$

$$\left(\frac{a+b}{a}\right)x = 1$$

$$x = \frac{a}{a+b},$$

$$y = \frac{b}{a}x = \left(\frac{b}{a}\right)\left(\frac{a}{a+b}\right) = \frac{b}{a+b}.$$

Edge Method for Dividing the Square into *n*ths

To use this method, first locate a key point on the left edge. Make a fold using that point, and the division is found on the top edge.

Example: Divide into 7ths.

We begin with two numbers whose sum is 7 and one of the numbers is the greatest power of 2 less than 7. That power of 2 is 4. So one number is 4, and the other is 3.

Make a fraction of the two numbers. This would be 3/4. Find height 3/4 on the left edge. Bring the bottom right corner to the top edge and the bottom edge to the 3/4 mark on the left. The top edge is divided into 7ths by the bottom right corner.

1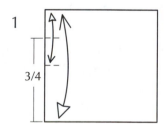

Make small marks by folding and unfolding in quarters.

2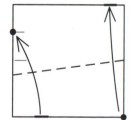

Bring the bottom right corner to the top edge and the bottom edge to the 3/4 mark on the left.

3

Fold and unfold.

4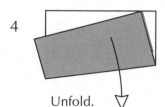

Unfold.

5

The 1/7 mark is on the top edge.

Edge method for dividing the square into _n_ths.

Given _n_, find a_T and a_B so $n = a_T + a_B$ where $a_B = 2^m$ is the largest power of $2 < n$.

Example: $n = 7$ then $a_B = 4$, $a_T = 3$.
$n = 11$ then $a_B = 8$, $a_T = 3$.

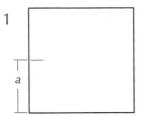

1

Find the location
of $a = a_T / a_B$ on
the left edge.

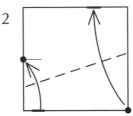

2

Bring the lower right corner to
the top edge and the bottom
edge to the left landmark.

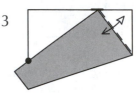

3

Fold and unfold.

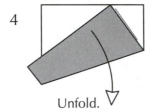

4

Unfold.

5

$$b = \frac{1-a}{1+a}$$

Edge Method for Dividing the Square into n^2 Parts

In this method, locate a key point on the left edge. Make a fold using the key point, and divisions are found on the left and right edges.

Example: Divide into 25ths.

We begin with the square root of 25, which is 5. Find two numbers whose sum is 5 and one of the numbers is the greatest power of 2 less than 5. That power of 2 is 4. So one number is 4, and the other is 1.

Make a fraction of the two numbers. This would be 1/4. Find height 1/4 on the left edge. Bring the bottom right corner to the top edge and the bottom edge to the 1/4 mark on the left. The left and right edges are divided into 25ths by the crease.

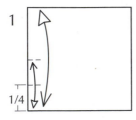

1

Make small marks by
folding and unfolding
in quarters.

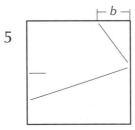

2

Bring the bottom right
corner to the top edge
and the bottom edge to
the 1/4 mark on the left.

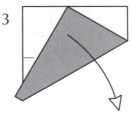

3

Unfold.

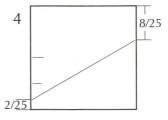

4

Opposite sides are divided
into 2/25 and 8/25.

Edge method for dividing the square into n^2 parts.

Given n^2, find a_T and a_B so $n = a_T + a_B$ where $a_B = 2^m$ is the largest power of $2 < n$.

Example: $n^2 = 25$ then $a_B = 4, a_T = 1$.
             $n^2 = 81$ then $a_B = 8, a_T = 1$.

1

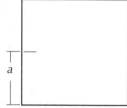

Find the location of $a = a_T/a_B$ on the left edge.

2

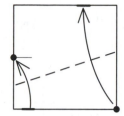

Bring the lower right corner to the top edge and the bottom edge to the left crease.

3

Unfold.

4

$$c = \frac{2a}{(1+a)^2}$$

$$d = \frac{2a^2}{(1+a)^2}$$

Derivation of the edge method.

1. Given b, find c.

1

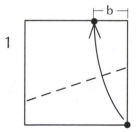

Given landmark b, fold the lower right corner to b.

2

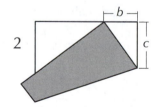

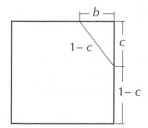

The hypotenuse of the triangle $= 1 - c$, so

$$b^2 + c^2 = (1-c)^2$$
$$b^2 + c^2 = 1 - 2c + c^2$$
$$b^2 = 1 - 2c$$
$$c = (1 - b^2)/2$$

b	c
1/2	3/8
1/4	9/32
3/4	7/32
$1/\sqrt{2}$	1/4
$\sqrt{3}/2$	1/8

Examples of b and c values for some easy landmarks.

2. Given *a*, find *b* and *c*.

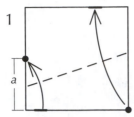

1

Bring the lower right corner to the top edge and the bottom edge to *a*.

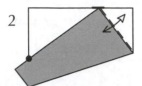

2

Fold and unfold.

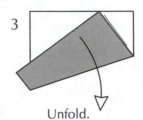

3

Unfold.

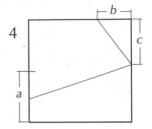

4

The dotted lines on the right represent the bottom two corners when the first fold is performed. The angle in the upper left triangle is $180° − 90° − β = α$. The two triangles whose sides meet the upper edge are similar.

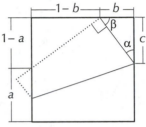

Label segments and angles.

Because the triangles are similar,

$$\frac{1-a}{1-b} = \frac{b}{c}.$$

Since $c = (1-b^2)/2$,

$$\frac{1-a}{1-b} = \frac{2b}{1-b^2}$$

$$1-a = \frac{2b(1-b)}{1-b^2}$$

$$= \frac{2b(1-b)}{(1-b)(1+b)}$$

$$= \frac{2b}{1+b},$$

so

$$a = \frac{1-b}{1+b}.$$

And solving for *b*,

$$(1+b)a = 1-b$$

$$a + ab + b = 1$$

$$b(a+1) = 1-a$$

$$b = \frac{1-a}{1+a}.$$

Since $c = (1-b^2)/2$ and

$$b = \frac{1-a}{1+a}$$

then, after some calculation,

$$c = \frac{2a}{(1+a)^2}.$$

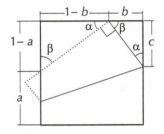

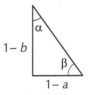

3. Given *b*, find *d*.

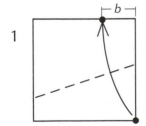

1

Given landmark *b*, fold the lower right corner to *b*.

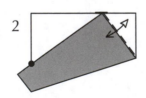

2

Fold and unfold. The dot locates *a*.

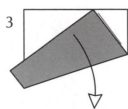

3

Unfold.

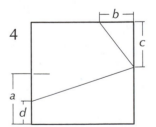

4

Solve for d:

$$d + e = a$$

Because of similar triangles

$$\frac{d}{e} = \frac{c}{1-c}$$

$$\frac{d}{a-d} = \frac{c}{1-c}$$

$$d(1-c) = c(a-d)$$

$$d - dc = ca - cd.$$

Thus

$$d = ca$$

$$= \left(\frac{1-b^2}{2}\right)\left(\frac{1-b}{1+b}\right)$$

$$= \frac{(1-b)(1+b)(1-b)}{2(1+b)}$$

$$= (1-b)^2/2.$$

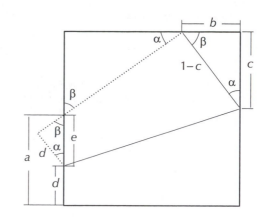

4. Given a, find b, c, and d.

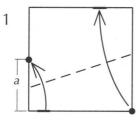

1

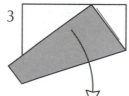

2 Fold and unfold.

3 Unfold.

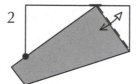

4

Bring the lower right corner to the top edge and the bottom edge to a.

Landmarks b, c, and d are determined by a.

Given a,

$$b = \frac{1-a}{1+a},$$

$$c = \frac{2a}{(1+a)^2}.$$

Given b,

$$c = (1-b^2)/2,$$

$$d = (1-b)^2/2.$$

So,

$$d = \frac{\left(1 - \frac{1-a}{1+a}\right)^2}{2}$$

$$= \frac{\left(\frac{1+a-1+a}{1+a}\right)^2}{2}$$

$$= \frac{\left(\frac{2a}{1+a}\right)^2}{2}$$

$$= \frac{2a^2}{(1+a)^2}.$$

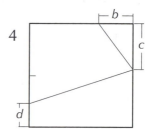

Given a

$$b = \frac{1-a}{1+a}$$

$$c = \frac{2a}{(1+a)^2}$$

$$d = \frac{2a^2}{(1+a)^2}$$

a	b	c	d
1/2	1/3	4/9	2/9
1/4	3/5	8/25	2/25
3/4	1/7	24/49	18/49
1/8	7/9	16/81	2/81
3/8	5/11	48/121	18/121
5/8	3/13	80/169	50/169

Final formula and examples of some values.

Angles

When we design polyhedra, angles are needed. Here are folding sequences to form mainly integer angles. Some are key angles that generate others.

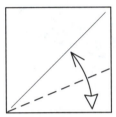

For a 1 × 1 square,

$$a = \tan(\alpha)$$

45° and 22.5°

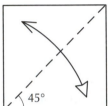

1

Fold along the diagonal.

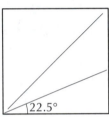

2

Fold and unfold
to the diagonal.

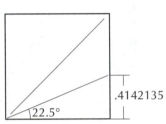

3

$$\tan(22.5°) \approx .4142135 \approx \sqrt{2} - 1$$

30°

1

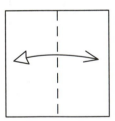

Fold and unfold.

2

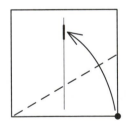

Fold the dot to the
center crease.

3

Unfold.

4

30°

tan(30°) ≈ .57735 ≈ √3/3

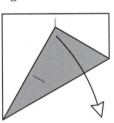

Note the dotted triangle from step 3.
The hypotenuse is 1, and the opposite
side is .5. Thus the angle is 30°.

15°

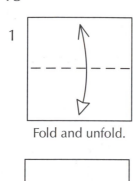

1 — Fold and unfold.

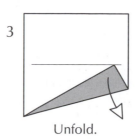

2 — Fold to the crease.

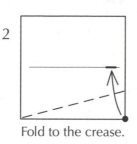

3 — Unfold.

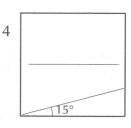

4

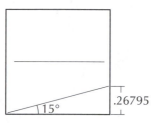

$\tan(15°) \approx .26795 \approx 2 - \sqrt{3}$

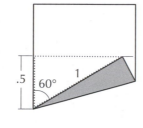

Note the dotted triangle from step 3. The hypotenuse is 1, and the adjacent side is .5. Thus the angle is 60°, making the angle in step 4 to be 30°/2 = 15°.

32°

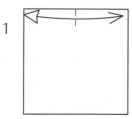

Since arctan(5/8) ≈ 32.00538°, the folding method for divisions of 1/8 can be used as a simple way to obtain 32°.

Folding method to make 32°:

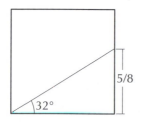

1 — Fold and unfold at the top.

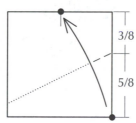

2 — Fold on the right.

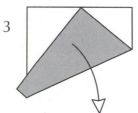

3 — Unfold.

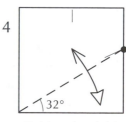

4 — Fold and unfold.

Here is a way to find more integer angles from 32°.

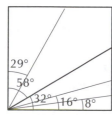

Given 32°, several other integer angles can be found. Note that 32° divided in half five times is 1°. So, in theory, all the integer angles can be found from it!

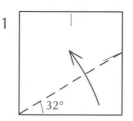

1

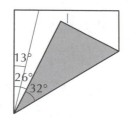

2

Here is a list of some angles that are easy to find given 32°:

1°, 2°, 4°, 8°, 13°, 16°, 22°, 26°, 29°.

14°

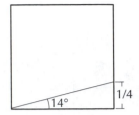

Since arctan(1/4) ≈ 14.0362°, a simple way to obtain 14° is to mark 1/4 on the right edge.

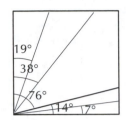

More integer angles can be found from 14°, including 28° since 2 × 14° = 28°. These angles can be found given 14°:

7°, 19°, 28°, 38°, 76°.

24°

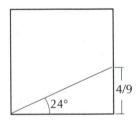

Since arctan(4/9) ≈ 23.96°, the folding method for divisions of 1/9 can be used as a simple way to obtain 24°.

Folding method to make 24°:

1

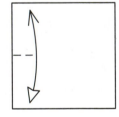

Fold and unfold on the left.

2

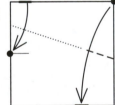

Bring the upper right corner to the bottom edge and the top edge to the left center. Crease on the right.

3

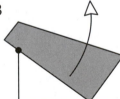

Unfold.

4

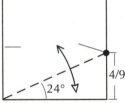

Fold and unfold at 24°.

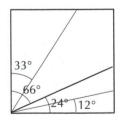

Here is a list of some angles that are easy to find given 24°:

3°, 6°, 12°, 21°, 33°, 39°, 42°, 48°.

25°

Folding method to make 25°:

1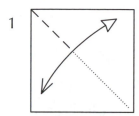
Fold and unfold by the top.

2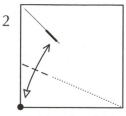
Fold and unfold on the left.

3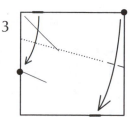
Bring the top edge to the left dot and the top right corner to the bottom edge. Crease on the right.

4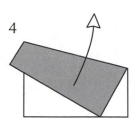
Unfold.

5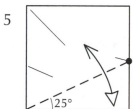
Fold and unfold.

6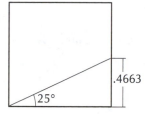
tan(25°) ≈ .4663

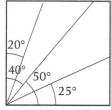

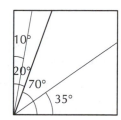

More integer angles can be found given 25°. Here are of some of them:

5°, 10°, 20°, 35°, 40°, 50°.

36°

Folding method to make 36°:

1
Fold and unfold by the top.

2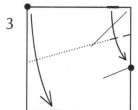
Fold and unfold on the right.

3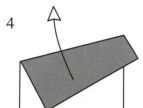
Bring the top edge to the right dot and the top left corner to the bottom edge. Crease on the right.

4
Unfold.

5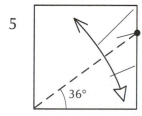
Fold and unfold.

6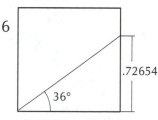
tan(36°) ≈ .72654

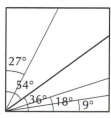

More integer angles can be found given 36°. Here are of some of them:

9°, 18°, 27°, 54°.

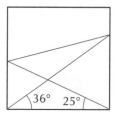

The folding methods for finding 25° and 36° are almost the same.

Angles through the Center

The crease patterns of many polyhedra have a line going through the center of the square paper at some angle.

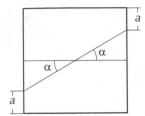

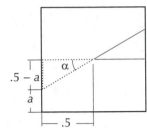

In the dotted triangle

$$\tan(\alpha) = (.5 - a)/.5,$$

so

$$a = .5(1 - \tan(\alpha))$$

In general, given angle α, calculate a.
Locate the landmarks on opposite edges.

Suppose you can fold an angle from the corner of the square and now want to fold that angle through the center. Here is a method.

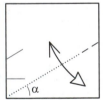

1

Fold and unfold along the known angle but only crease on the right.

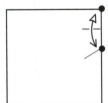

2

Fold and unfold on the right.

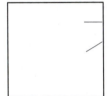

3

Rotate 180°.

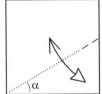

4

Repeat steps 1–2.

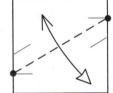

5

Fold and unfold.

6

The line goes through the center at angle α.

30° through the Center

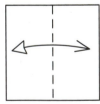

1

Fold and unfold.

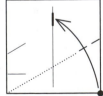

2

Fold the dot to the vertical line and crease on the right.

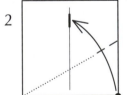

3

Unfold.

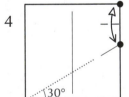

4

Fold and unfold on the right. Rotate 180°.

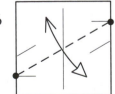

5

Repeat steps 2–4.

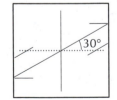

6

Fold and unfold.

7

The lines goes through the center at 30°.

Design Method Examples

Here are some examples to illustrate the process behind designing origami polyhedra. We follow the steps outlined at the beginning of "Design Factors and Techniques" and implement tools discussed there and in "Math and Design." In the end, I hope that I have given some insight into design concepts, as there is still so much to be discovered and enjoyed.

Design Method for the Hexagonal Dipyramid in a Sphere

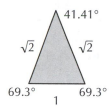

Hexagonal Dipyramid
in a Sphere

The lengths of the sides are proportional to $1, \sqrt{2}, \sqrt{2}$.

This solid is composed of twelve isosceles triangles. Some math is require to find the angles and lengths of the sides of each triangular face. The formulas are given in "Dipyramid Design" (page 173).

Layouts

Here are a few.

Radial

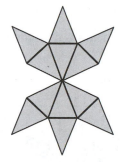

Even/odd

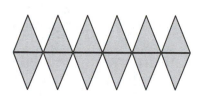

Even/odd

Even/odd, butterfly layout

Odd

Even/odd

Crease Patterns

Here are just a few possibilities.

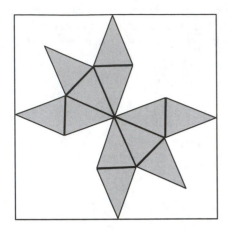 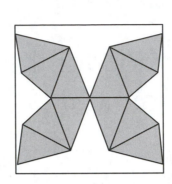 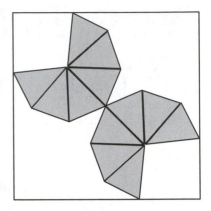

In each of these crease patterns, the size of the triangular faces are the same. Thus, the smaller the square enclosing the crease pattern, the larger the finished model.

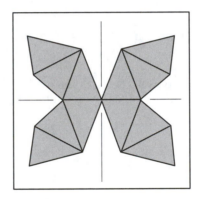

After experimenting with these and other crease patterns, I discovered that this one is the best. It has good symmetry and locks together well.

Landmarks

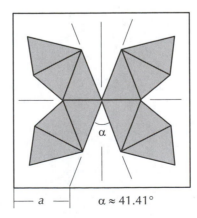

$\alpha \approx 41.41°$

Given the angle of 41.41°, length a is calculated to be .3110. Robert Lang's ReferenceFinder software is used to find a short folding procedure to locate point a.

Design Method for the Sunken Tetrahedron Using Math

Designing the sunken tetrahedron requires a bit of math. First, we find the proportions of the sides of each triangular face. We then develop a layout with landmarks.

    The sunken tetrahedron is composed of 12 isosceles triangles. The proportions of the sides of the triangle are to be found.

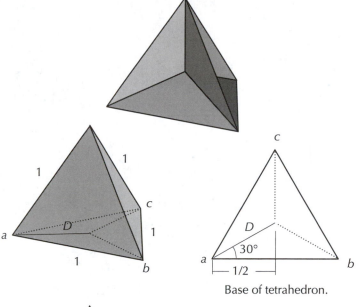

1. Begin with a tetrahedron where the edges are all 1 unit in length. Find the length of line D, on the base from a vertex to the center.

   To solve for D:

   $$\cos(30°) = \frac{1/2}{D}$$

   $$D = \frac{1}{2\cos(30°)} = \frac{1}{\sqrt{3}}$$

   Base of tetrahedron.

2. Find height H.

   $$1 = H^2 + \left(\frac{1}{\sqrt{3}}\right)^2$$

   $$H = \sqrt{\frac{2}{3}}$$

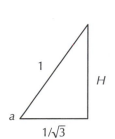

3. Find length F, which represents one of the sides of the triangles of the sunken tetrahedron. Since triangles T_1 and T_2 are similar:

   $$\frac{F}{1/2} = \frac{1}{\sqrt{2/3}}$$

   $$F = \frac{\sqrt{3}}{2\sqrt{2}}$$

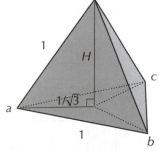
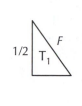
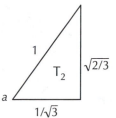

4. The proportions are found, as shown on the right.

   This shows the proportions of the sides. To make work for the values easier, the numbers are scaled so that the proportions are $\sqrt{3}$, $\sqrt{3}$, and $2\sqrt{2}$.

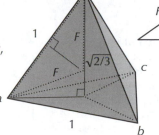
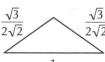

Layouts

Now that the proportions of the triangles are known, a layout can be made. There are several possibilities. Here are a few.

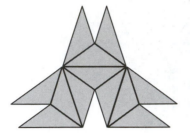

Even symmetry.

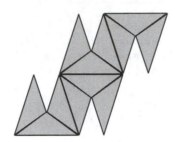

Odd symmetry.

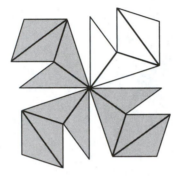

3/4 square symmetry.

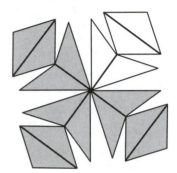

3/4 square symmetry.

Crease Patterns

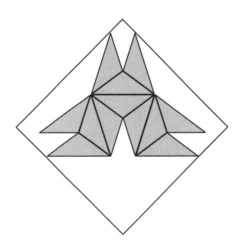

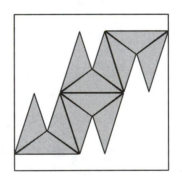

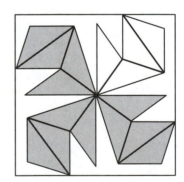

After experimenting with these crease patterns, this one was chosen. It yields the largest model that has the fewest steps and holds together well.

If it does not maximize the size of the model, then it is very close to it and appears to maximize with respect to its simplicity. It uses 3/4 square symmetry.

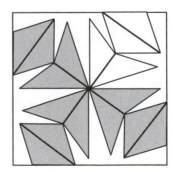

Landmarks

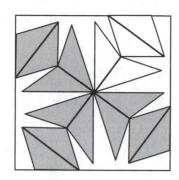

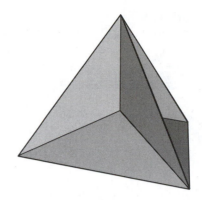

Given the crease pattern, the landmarks are to be found.

1. Find α.

$$\tan(\alpha) = 1/\sqrt{2}$$
$$\alpha = \arctan(1/\sqrt{2})$$
$$= 35.26439°$$

Triangular face.

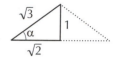

2. Find a. (This crease is used in the folding.)

$$\theta = (90° - 2\alpha)/2$$
$$= 45° - \alpha$$
$$\tan(\theta) = b/.5 = 2b$$
$$b = .5\tan(\theta)$$

$$a = .5 - b$$
$$= .5(1 - \tan(\theta))$$
$$= .4142135$$
$$= \sqrt{2} - 1$$
$$= \tan(22.5°)$$

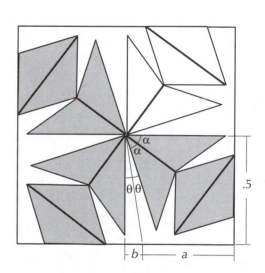

3. Find c.

$$\phi = 90° - 2\alpha$$
$$\tan(\phi) = d/.5 = 2d$$
$$d = .5\tan(\phi)$$

$$c = .5 - d$$
$$= .5(1 - \tan(\phi))$$
$$= .3232233$$
$$= (4 - \sqrt{2})/8$$

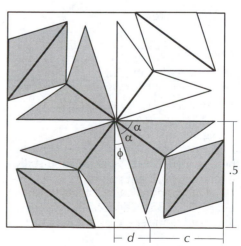

4. Find e.

$$\tan(\alpha) = f/.5$$
$$f = .5\tan(\alpha)$$

$$e = .5 - f$$
$$= .5(1 - \tan(\alpha))$$
$$= .1464466$$
$$= (2 - \sqrt{2})/4$$

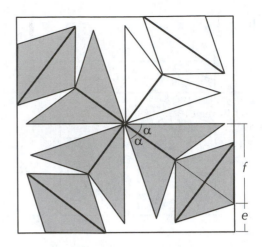

5. Find g, first find k.

Proportion

$$k/h = .5/(h + 2j)$$

and

$$h = \sqrt{3}j$$
$$j = h/\sqrt{3}$$

so

$$h + 2j = h + (2/\sqrt{3})h$$
$$= h(1 + 2/\sqrt{3})$$
$$k = \frac{h}{2h(1 + 2/\sqrt{3})}$$
$$= \frac{1}{2(1 + 2/\sqrt{3})}$$
$$g = .5 - k$$
$$= .5\left(1 - \frac{1}{1 + 2/\sqrt{3}}\right)$$
$$= .2679491$$
$$= \tan(15°).$$

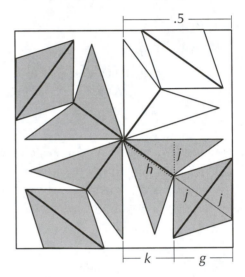

The landmarks are

$$a = .4142135 = \tan(22.5°)$$

$$c = .3232233 = (4 - \sqrt{2})/8$$

$$e = .1464466 = (2 - \sqrt{2})/4$$

$$g = .2679491 = \tan(15°)$$

Of these, a and g are easy to find and sufficient for completing the folding pattern.

Find landmark a by folding the angle 22.5°, since $a = .4142135 = \tan(22.5°)$.

Find landmark g by folding the angle 15°, since $g = .2679491 = \tan(15°)$.

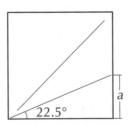

$a = .4142135 = \tan(22.5°)$

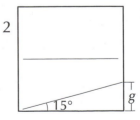

Fold and unfold. $g = .2679491 = \tan(15°)$

Polygons

Since polygons are the faces of polyhedra, we present how to fold certain polygons before we move into folding directions for polyhedra.

Many of the polyhedra in this collection have faces that are regular polygons, which are polygons with congruent angles and sides. Each interior angle between two sides of a regular n-gon $= (n - 2)180°/n$.

Regular Polygons

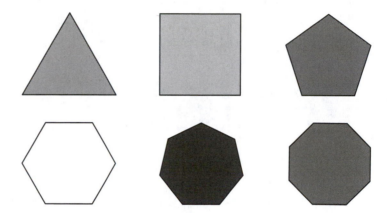

n	Polygon	Angle at vertex
3	Triangle	60°
4	Square	90°
5	Pentagon	108°
6	Hexagon	120°
7	Heptagon	128.57°
8	Octagon	135°
9	Nonagon	140°
19	Decagon	144°
11	Hendecagon	147.27°
12	Dodecagon	15°

One of the easiest ways to create a regular polygon in a square is by the *book-fold symmetry* method. In this method, start with a vertex centered on the top of the square. Create lines that all radiate from that top-center point and that divide the edge into n equal angles. Let the top-center point be the top vertex of a regular n-gon; all of the vertices will lie on the lines. All the lines will meet the vertices of the n-gon no matter the size of the n-gon.

The angle between any two adjacent lines radiating from the top center of the square is $\alpha = 180°/n$.

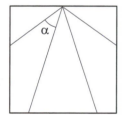

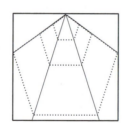

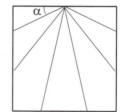

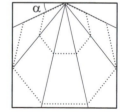

$\alpha = 180°/5 = 36°$

The lines meet the vertices of a pentagon.

For a heptagon, $\alpha = 180°/7 \approx 25.7°$.

Rectangles are also used in polyhedra design. Folding directions will be given for rectangles based on three famous ratios.

Rectangles

Rectangle	Dimensions
Silver Rectangle	$1 \times \sqrt{2}$ (≈ 1.414)
Bronze Rectangle	$1 \times \sqrt{3}$ (≈ 1.732)
Golden Rectangle	$1 \times (1 + \sqrt{5})/2$ ($= \phi \approx 1.618$)

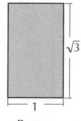

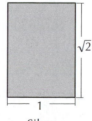

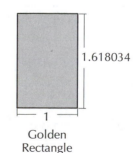

$\sqrt{2}$

$\sqrt{3}$

1.618034

1

1

1

Silver Rectangle

Bronze Rectangle

Golden Rectangle

Equilateral Triangle

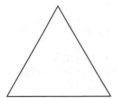

Equilateral Triangle

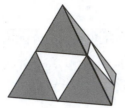

Tetrahedron of Triangles

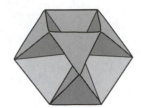

Octahemioctahedron

Icosahedron

Equilateral triangles are used in several polyhedra. The tetrahedron of triangles, octahemioctahedron, and icosahedron are composed entirely of them.

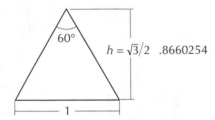

This shape has three sides with interior angles of 60°. For sides of length 1, the height is $\sqrt{3}/2$.

$h = \sqrt{3}/2$ .8660254

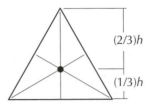

$(2/3)h$

$(1/3)h$

The bisectors—lines that divide each angle in half—divide the height into thirds where they intersect.

Triangles placed in a layout for polyhedra come in different orientations. Two methods for folding triangles will be given: book-fold and diagonal symmetry.

Book-fold Symmetry

The triangle can be oriented at different heights in the square. Here is the layout following the book-fold symmetry method, with the top vertex meeting the top of the square.

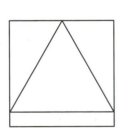

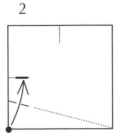

.5

60°

60°

30°

$\sqrt{3}/2 \approx .8660254$

$1 - \sqrt{3}/2 \approx .1339746$

1

Landmarks.

Folding method:

1

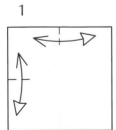

Fold and unfold in half on two edges.

2

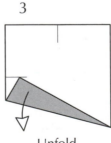

Bring the dot to the crease. Crease on the left.

3

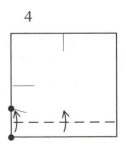

Unfold.

4

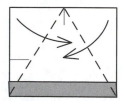 5

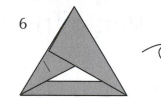 6

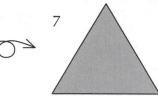 7

Equilateral Triangle

Diagonal Symmetry

Here, the height of the triangle is along the diagonal fold instead of the book fold. This layout yields the largest triangle in a square.

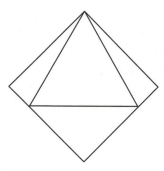

The length of the side of the triangle = 1/cos(15°) ≈ 1.0352762. For the book-fold method, the length is 1.

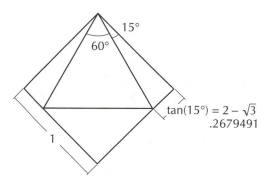

$\tan(15°) = 2 - \sqrt{3}$
.2679491

Landmarks.

Folding method:

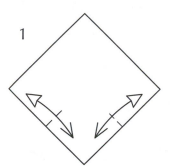 1

Fold and unfold in half on two edges.

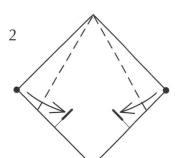 2

Bring the dots to the creases.

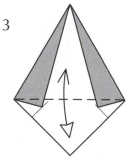 3

Fold and unfold.

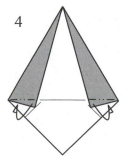 4

Fold behind.

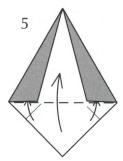 5

Fold inside.

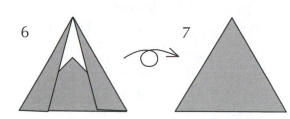 6

7

Equilateral Triangle

Pentagon

Pentagon

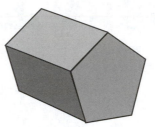

Dodecahedron

Golden Pentagonal Prism

Golden Pentagonal
Antiprism

Pentagonal faces are on the dodecahedron, golden pentagonal prism, and golden pentagonal antiprism.

Some calculations are required to determine the landmarks.

1. Find width w.
   Let the length of each side = 1.
   Find α.
   $$2\alpha + 108° = 180°$$
   $$\alpha = 36°$$
   By the Law of Sines
   $$\frac{w}{\sin(108°)} = \frac{1}{\sin(36°)}$$
   $$w = \sin(108°)/\sin(36°)$$
   $$= 1.618034$$
   $$= (1+\sqrt{5})/2$$
   $$= \phi, \text{ the golden mean.}$$

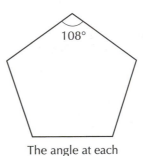

The angle at each
vertex = 108°.

2. Find height h.
   First find angle β.
   $$\alpha + \beta = 108°/2$$
   Since $\alpha = 36°$ then $\beta = 18°$.
   $$\tan(18°) = .5/h$$
   $$h = 1/(2\tan(18°))$$
   $$\approx 1.5388418$$

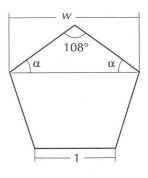

3. Find the ratio of h_b to h, where h_b is
   the height of the bottom section.
   $$\beta = 108° - 90° = 18°$$
   $$h_b = \cos(18°) \approx .9510565$$
   $$h_b/h = .95105651/.5388418$$
   $$= .618034$$
   $$= \phi - 1$$

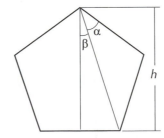

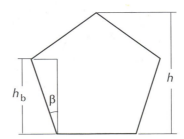

4. Scale the pentagon so $h_b = 1$.

Then
$$h = 1.618034 = \phi$$
and
$$h_t = .618034 = \phi - 1$$

where h_t is the height of the top section.

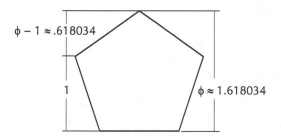

$\phi - 1 \approx .618034$

1

$\phi \approx 1.618034$

The pentagon can be folded with book-fold or diagonal symmetry.

Book-fold Symmetry

The pentagon can be oriented at different heights in the square. In this layout, the pentagon meets the top of the square.

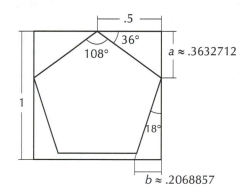

.5

36°

108°

$a \approx .3632712$

1

18°

$b \approx .2068857$

Landmarks:
$$a = .5\tan(36°) \approx .3632712$$
$$b = (1 - a)\tan(18°) \approx .2068857$$

Folding method:

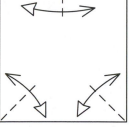

1

Fold and unfold in half
on the corners and edge.

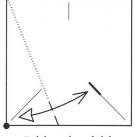

2

Fold and unfold
on the bottom.

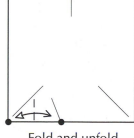

3

Fold and unfold
on the bottom.

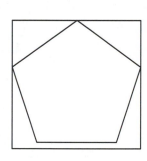

4

Repeat steps 2–3
on the right.

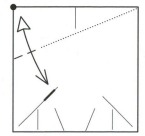

5

Fold and unfold
on the left.

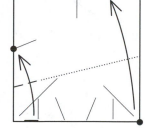

6

Bring the bottom edge to the dot on
the left and the bottom right corner
to the top edge. Crease on the left.

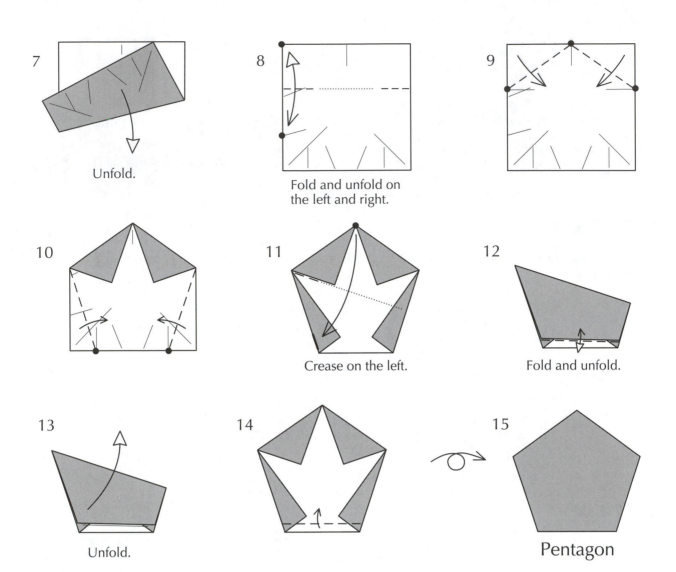

7   Unfold.

8   Fold and unfold on the left and right.

9

10

11   Crease on the left.

12   Fold and unfold.

13   Unfold.

14

15   Pentagon

Diagonal Symmetry

This layout yields the largest pentagon in a square.

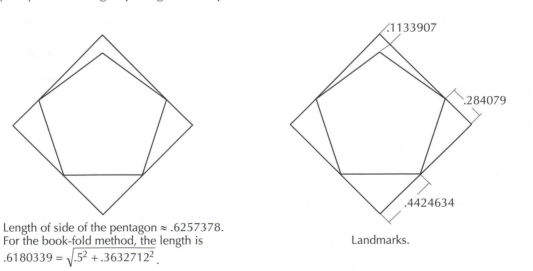

Length of side of the pentagon $\approx .6257378$.
For the book-fold method, the length is
$.6180339 = \sqrt{.5^2 + .3632712^2}$.

.1133907

.284079

.4424634

Landmarks.

Folding method:

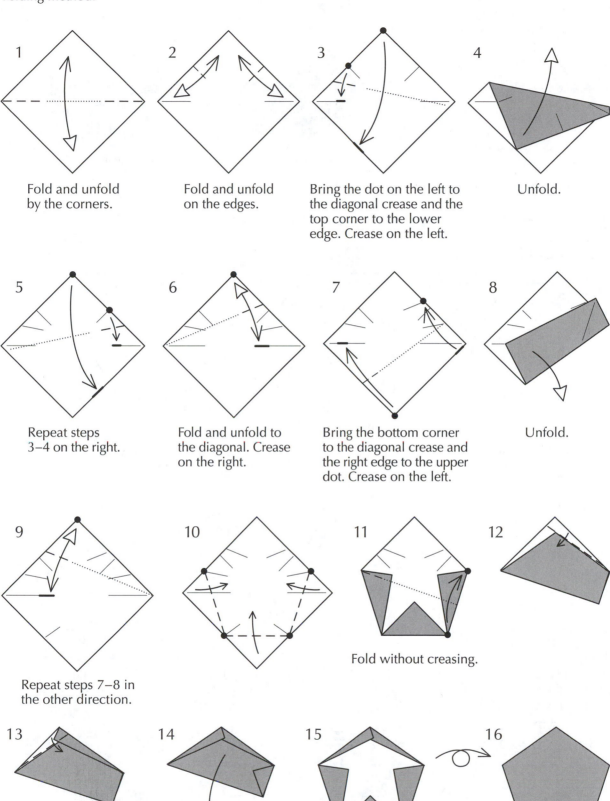

1 Fold and unfold by the corners.

2 Fold and unfold on the edges.

3 Bring the dot on the left to the diagonal crease and the top corner to the lower edge. Crease on the left.

4 Unfold.

5 Repeat steps 3–4 on the right.

6 Fold and unfold to the diagonal. Crease on the right.

7 Bring the bottom corner to the diagonal crease and the right edge to the upper dot. Crease on the left.

8 Unfold.

9 Repeat steps 7–8 in the other direction.

10

11 Fold without creasing.

12

13

14 Unfold.

15

16

Pentagon

Hexagon

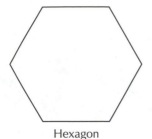 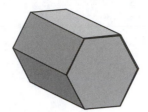 

Hexagon   Hexagonal Prism   Hexagonal Antiprism

Several polyhedra have hexagonal faces. These include the hexagonal prism and hexagonal antiprism.

The hexagon can be folded using book-fold and diagonal symmetry.

Book-fold Symmetry

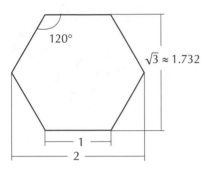

$120°$

$\sqrt{3} \approx 1.732$

$\frac{1}{2}$

2

The angle at each vertex = 120°.
Let length of each side = 1.
Then width = 2, height = $\sqrt{3}$.

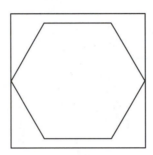

For this layout, the hexagon is placed in the center.

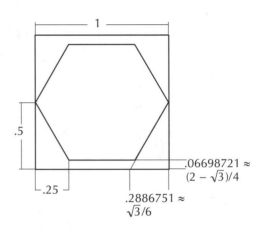

1

$.5$

$.25$

$.06698721 \approx (2 - \sqrt{3})/4$

$.2886751 \approx \sqrt{3}/6$

Folding method:

1

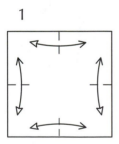

Fold and unfold in half on the edges.

2

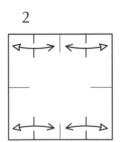

Fold and unfold.

3

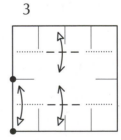

Fold and unfold in the center.

4

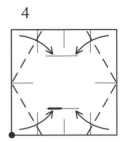

Fold the corners to the creases.

5

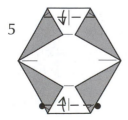

6

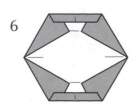

7

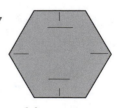

Hexagon

Diagonal Symmetry

This layout yields the largest hexagon in a square.

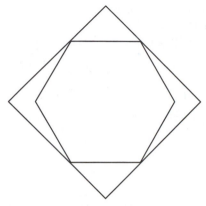

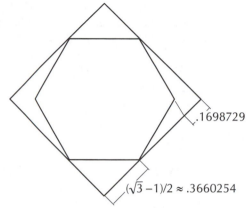

Length of side of hexagon ≈ .517638.
For the book-fold method, the length is .5.

Landmarks.

.1698729

$(\sqrt{3} - 1)/2 \approx .3660254$

Folding method:

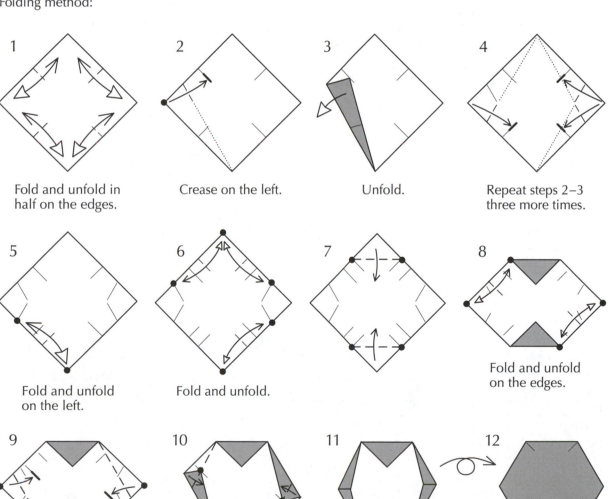

1 Fold and unfold in half on the edges.

2 Crease on the left.

3 Unfold.

4 Repeat steps 2–3 three more times.

5 Fold and unfold on the left.

6 Fold and unfold.

7

8 Fold and unfold on the edges.

9

10

11

12

Hexagon

Heptagon

There are polyhedra that have heptagonal sides. This includes a heptagonal pyramid and antiprism. In this collection, there are some heptagonal dipyramids, but their faces are triangles.

The angle at each vertex ≈ 128.57°.

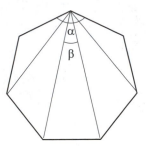

$\beta = 180°/7 \approx 25.71°$
$\alpha = 5\beta \approx 128.57°$

Some calculations are required to determine the landmarks.

1. Find width w.
   Let length of each side of the heptagon = 1.
   First find length d. By the Law of Sines:
   $$d/\sin(5\beta) = 1/\sin(\beta)$$
   $$d = \sin(5\beta)/\sin(\beta)$$
   $$\approx 1.80194$$
   Now find w, again using the Law of Sines:
   $$w/\sin(3\beta) = d/\sin(2\beta)$$
   $$w = (d)\sin(3\beta)/\sin(2\beta)$$
   $$\approx 2.24698$$

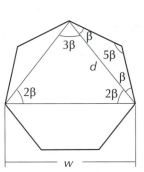

2. Find height h.
   $$\tan(\beta/2) = 1/(2h)$$
   $$h = 1/(2\tan(\beta/2))$$
   $$\approx 2.190643$$

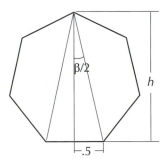

Folding directions will be given for the Heptagon using book-fold symmetry.

Book-fold Symmetry

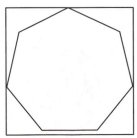

The heptagon meets
the top of the square.

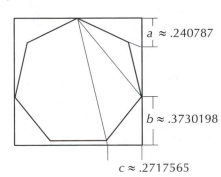

$a \approx .240787$

$b \approx .3730198$

$c \approx .2717565$

Landmarks:
$a = (.5)\tan(180°/7) \approx .240787$
$b = 1 - (.5)\tan(360°/7) \approx .3730198$
$c = .5 - \tan(180°/14) \approx .2717565$

Folding method:

1

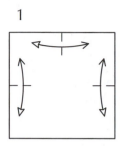

Fold and unfold in half on three edges.

2

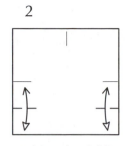

Fold and unfold.

3

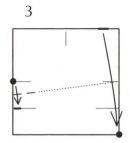

Bring the left edge to the crease on the left and the top edge to the bottom right corner. Crease on the left.

4

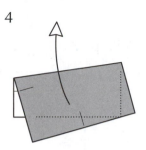

Unfold.

5

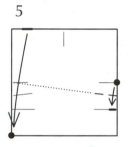

Repeat steps 3–4 in the other direction.

6

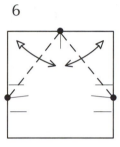

Fold and unfold creasing lightly.

7

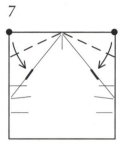

8

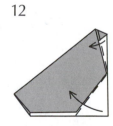

Crease on the left.

9

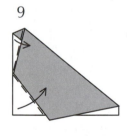

10

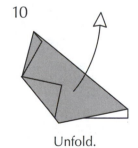

Unfold.

11

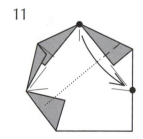

12

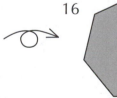

13

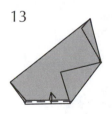

14

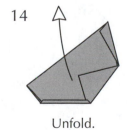

Unfold.

15

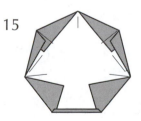

16

Heptagon

Octagon

Some polyhedra have octagonal sides, including an octagonal pyramid and antiprism.

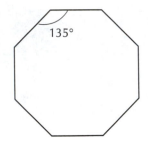

The angle at each vertex = 135°.

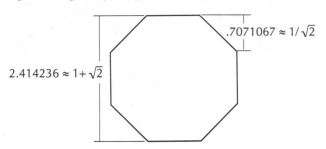

Let the length of each side = 1. The height of the octagon = 2.414236 = $1 + \sqrt{2}$.

Folding method:

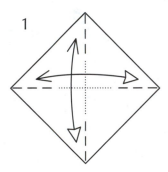

1

Fold and unfold
by the corners.

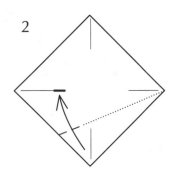

2

Crease on the left.

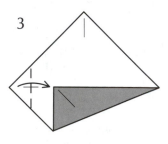

3

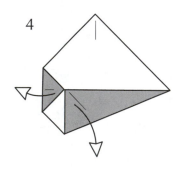

4

Unfold and rotate 90°.

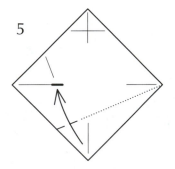

5

Repeat steps 2–4
three more times.

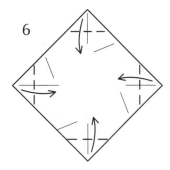

6

Refold.

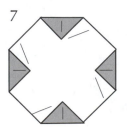

7

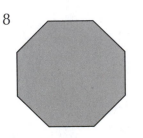

8

Octagon

Silver Rectangle

The silver rectangle has sides proportional to $1 \times \sqrt{2}$. This is the same as $\sqrt{2}/2 \times 1$, or $.7071 \times 1$. When folded in half, it keeps the same ratio. The European paper formats A4, A5, and A6 come in this proportion.

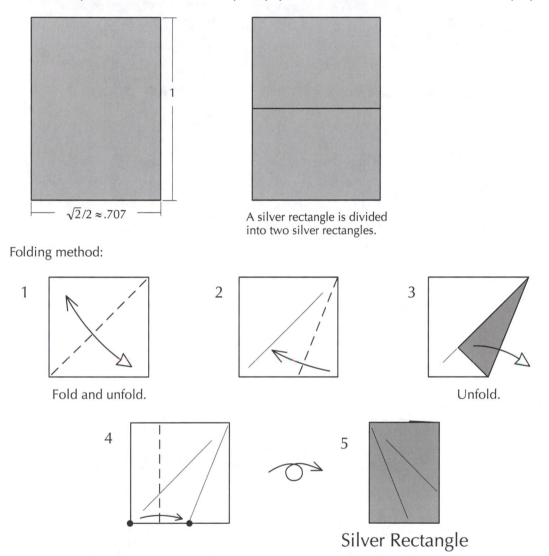

A silver rectangle is divided into two silver rectangles.

Folding method:

1   Fold and unfold.

2

3   Unfold.

4

5   Silver Rectangle

The silver rectangle is associated with $\sqrt{2}$ and 45° lines. The Tall Triangular Antiprism has faces that are 45° isosceles triangles.

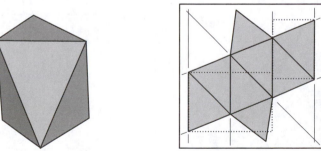

Tall Triangular Antiprism with layout and some silver rectangles marked with dotted lines.

Bronze Rectangle

The bronze rectangle has sides proportional to $1 \times \sqrt{3}$. This is the same as $\sqrt{3}/3 \times 1$, or $.57735 \times 1$. It is associated with equilateral triangles.

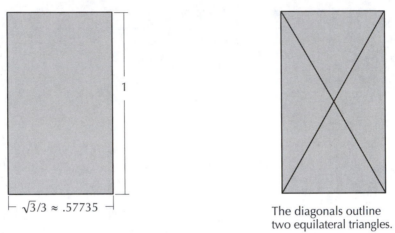

The diagonals outline two equilateral triangles.

Folding method:

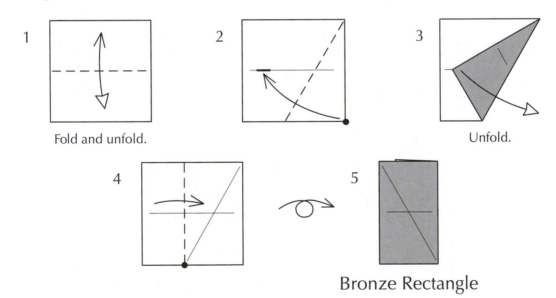

1. Fold and unfold.

2.

3. Unfold.

4.

5.

Bronze Rectangle

The bronze rectangle is associated with $\sqrt{3}$ and 30° or 60° lines. The folds for the Tetrahedron use the bronze rectangle.

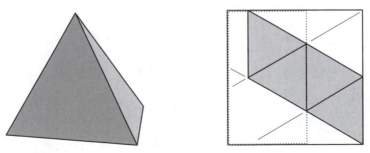

Tetrahedron with layout and a bronze rectangle marked with dotted lines.

Golden Rectangle

The golden rectangle has sides proportional to 1 × 1.618034. This is the same as .618034 × 1.

The name comes from the golden mean (phi = ϕ), $\phi \approx 1.618034 \approx (\sqrt{5} + 1)/2$. It is the solution to

$$x - 1 = 1/x$$

$$\phi - 1 \approx .618034$$

This number is associated with nature and beauty. We saw one example with the regular pentagon: if the length of each edge is 1, then the width is ϕ. There are many examples of things in nature that are divided by the golden mean, including the structure of the human body. If an arm is 1.618 units in length, then the lower arm is of length 1 and the upper arm is of length .618. The same applies to the distance from elbow to wrist and the length of a hand.

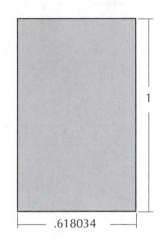

The golden rectangle divides into a square and a smaller golden rectangle.

Folding method:

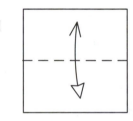

1

Fold and unfold.

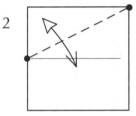

2

Fold and unfold.

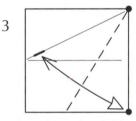

3

Fold and unfold.

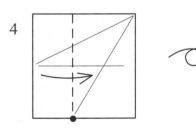

4

5

Golden Rectangle

The golden rectangle is associated with ϕ and pentagons. The folds for the Egyptian Pyramid use the golden rectangle.

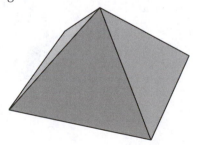
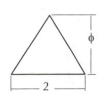
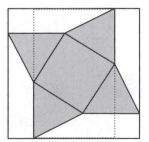

Egyptian Pyramid, one side, and layout with a golden rectangle marked with dotted lines.

Derivation of the Rectangles

Let us look into the calculations behind the folding directions for these three rectangles.

Silver Rectangle

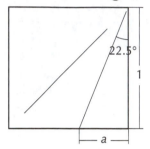 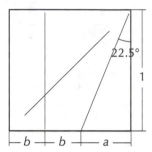 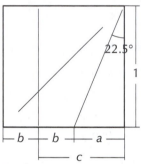 

Silver Rectangle

Find a.

$\tan(22.5°) = a/1$

$\quad a = \tan(22.5°)$

$\qquad \approx .4142135$

$\qquad = \sqrt{2} - 1$

Find b.

$2b + a = 1$

$\quad b = (1 - a)/2$

$\qquad \approx .2928932$

$\qquad = (2 - \sqrt{2})/2$

Find c.

$c = a + b$

$\quad \approx .7071067$

$\quad = 1/\sqrt{2}$

Bronze Rectangle

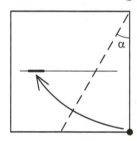 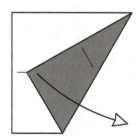 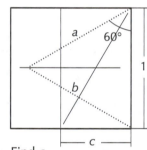 

Bronze Rectangle

Find angle α.

By setting up these folds, a (in the third drawing) has length 1. By symmetry, b also has length 1. An equilateral triangle is formed. Since α bisects 60°, $\alpha = 30°$.

Find c.

$\tan(30°) = c/1$

$\quad c = \tan(30°)$

$\qquad \approx .57735$

$\qquad = 1/\sqrt{3}$

Golden Rectangle

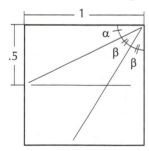 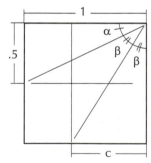

Golden Rectangle

Find angle α.

$\tan(\alpha) = 1/2$

$\quad \alpha = \arctan(1/2)$

$\qquad \approx 26.565°$

Find angle β.

$\alpha + 2\beta = 90$

$\quad \beta \approx 31.717°$

Find c.

$\tan(\beta) = c/1$

$\quad c = \tan(\beta)$

$\qquad \approx .618034$

$\qquad = (\sqrt{5} - 1)/2$

$\qquad = \phi - 1$

Part II

Platonic and Related Polyhedra

Tetrahedron Design

The tetrahedron is composed of four equilateral triangles.

Layout of the Tetrahedron

There are two main layouts. One is a triangle partitioned into four parts; the other is a band of four triangles.

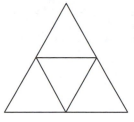
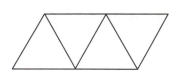

Crease Patterns

Let *s* be the length of a side of a triangle in a 1 × 1 square. A larger *s* value yields a larger tetrahedron from the same size paper.

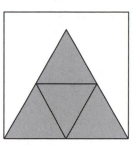

Even symmetry
s = .5

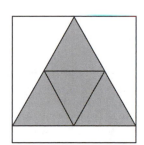

Even symmetry
s = .5

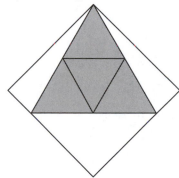

Even symmetry
s ≈ .5176
This makes the largest tetrahedron, but it does not have enough tab to hold.

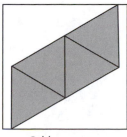

Odd symmetry
s = .5

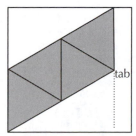

Not symmetric
s = .5
There is a tab on the right side. Without the tab, this would have odd symmetry. Used for the Tetrahedron (diagrammed).

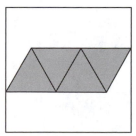

Odd symmetry
s = .4
Though this pattern would make a smaller tetrahedron, it could be used for effects such as a two-color version and is used for the Striped Tetrahedron.

Color Patterns

It is interesting to fold a tetrahedron with color patterns. Fold the square to show both sides of the paper and then choose a layout to create the pattern. Fold the tetrahedron from the layout.

Duo-Colored Tetrahedron

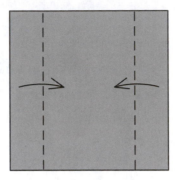

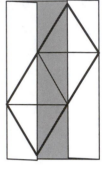

Fold the edges in so that a band of four triangles can be formed where each triangle shows both sides (colors) of the paper.

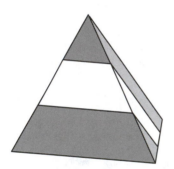

Striped Tetrahedron

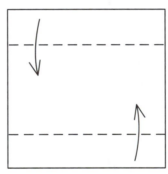

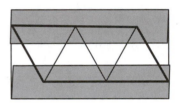

Fold the edges in so that the band of four triangles has a stripe going through it.

Tetrahedron Base

Several polyhedra can be formed using the tetrahedron as a base. These models have the same surface as a tetrahedron and thus share the same underlying structure.

Stellated Tetrahedron

Dimpled Truncated Tetrahedron

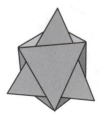

Tetrahedron in an Octahedron

(Sink the four corners of a tetrahedron in and then back out but not all the way.)

Tetrahedron

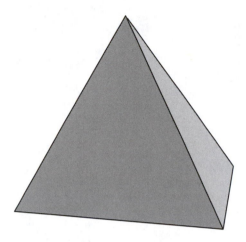

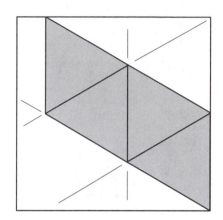

The tetrahedron, a Platonic solid composed of four triangles, is one of the simplest polyhedra. The crease pattern shows that it is constructed with a band of four triangles.

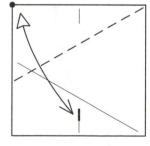

1

Fold and unfold on the top and bottom.

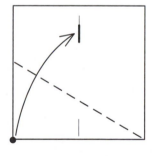

2

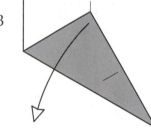

3

Unfold.

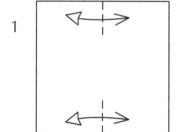

4

Fold and unfold.

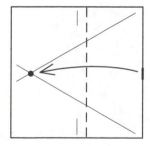

5

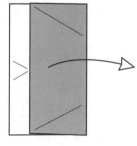

6

Unfold.

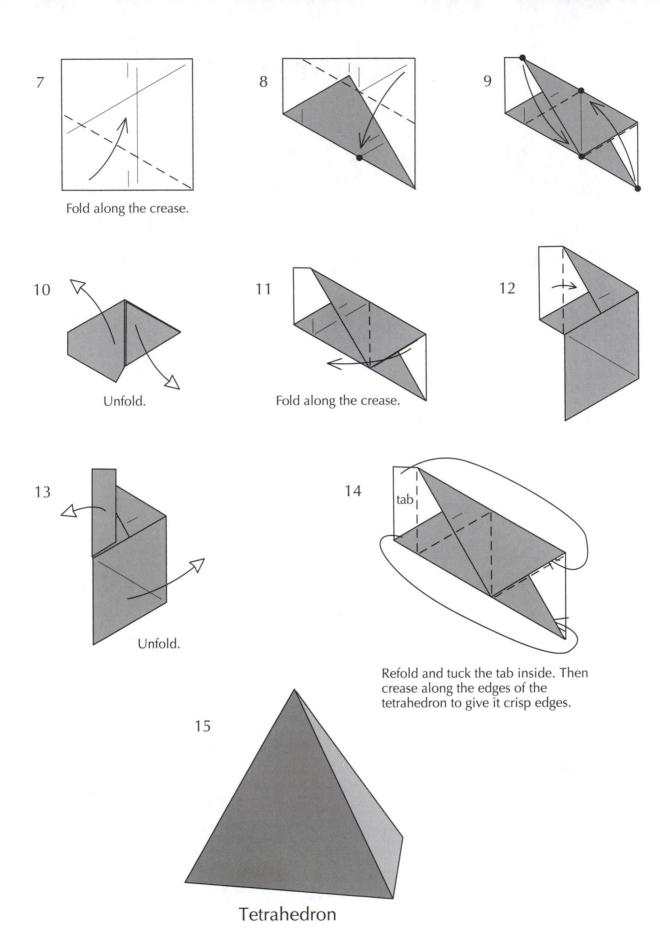

7 Fold along the crease.

8

9

10 Unfold.

11 Fold along the crease.

12

13 Unfold.

14 Refold and tuck the tab inside. Then crease along the edges of the tetrahedron to give it crisp edges.

tab

15

Tetrahedron

Duo-Colored Tetrahedron

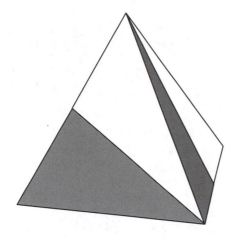

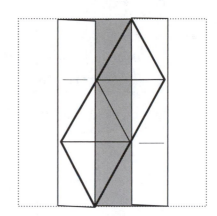

Each face of this tetrahedron shows both sides of the paper. Two opposite edges are folded a certain amount to show both sides of the paper. A band of four triangles with both colors gives the layout. The paper is divided into thirds.

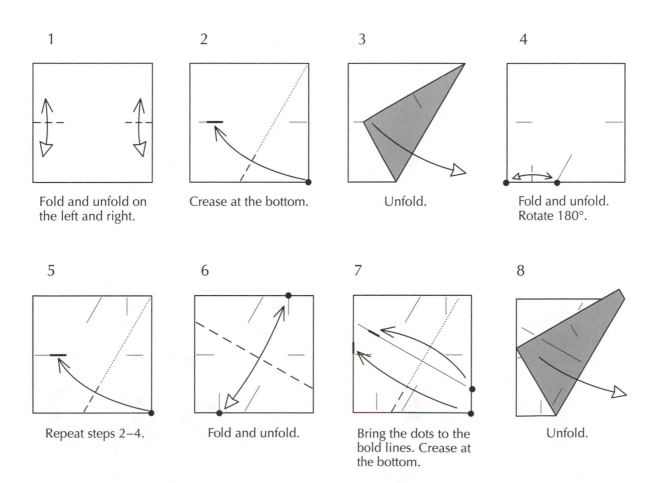

1

Fold and unfold on the left and right.

2

Crease at the bottom.

3

Unfold.

4

Fold and unfold.
Rotate 180°.

5

Repeat steps 2–4.

6

Fold and unfold.

7

Bring the dots to the bold lines. Crease at the bottom.

8

Unfold.

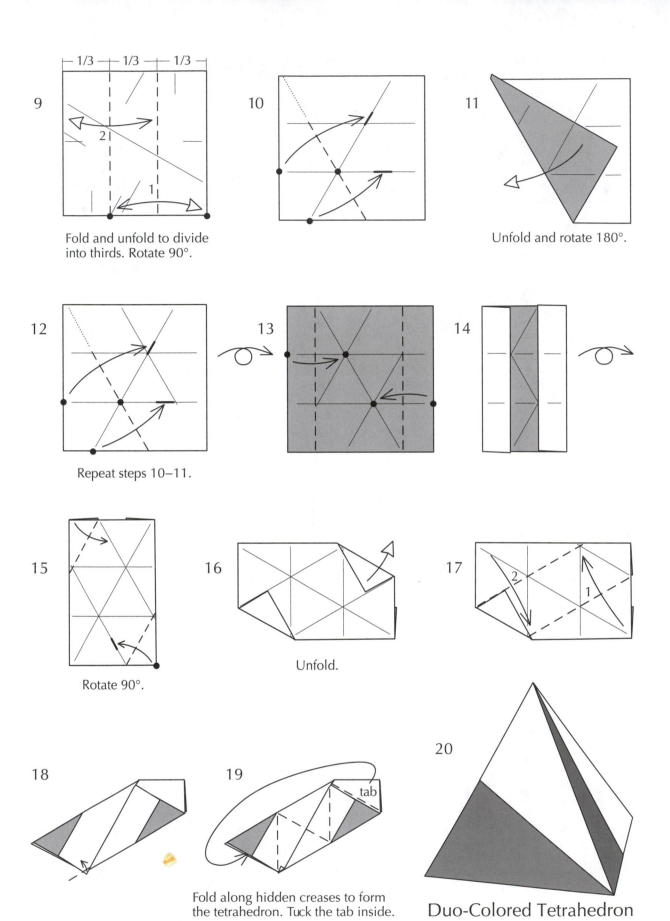

9

├─ 1/3 ─┼─ 1/3 ─┼─ 1/3 ─┤

2

1

Fold and unfold to divide
into thirds. Rotate 90°.

10

11

Unfold and rotate 180°.

12

Repeat steps 10–11.

13

14

15

Rotate 90°.

16

Unfold.

17

2   1

18

19

tab

Fold along hidden creases to form
the tetrahedron. Tuck the tab inside.

20

Duo-Colored Tetrahedron

Striped Tetrahedron

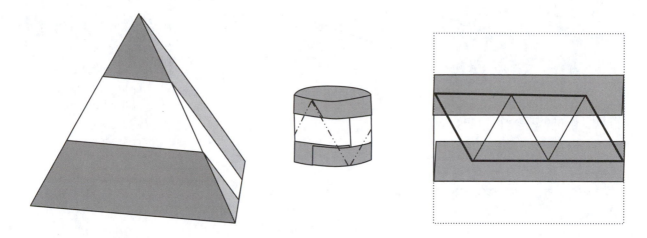

To fold this model, a striped cylinder is formed. With a few folds, the cylinder turns into the striped tetrahedron. The paper is divided into fifths.

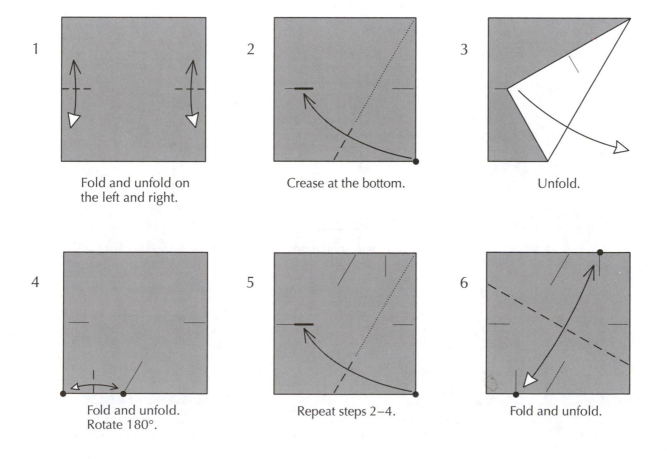

1. Fold and unfold on the left and right.

2. Crease at the bottom.

3. Unfold.

4. Fold and unfold. Rotate 180°.

5. Repeat steps 2–4.

6. Fold and unfold.

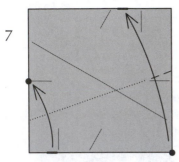

7

Bring the lower right corner to the top edge and the bottom edge to the left center. Crease on the right.

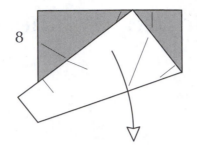

8

Unfold.

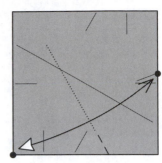

9

Fold and unfold at the bottom.

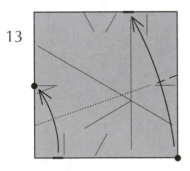

10

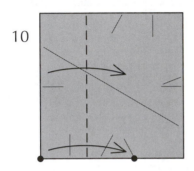

11

Fold and unfold along the crease.

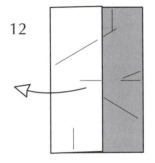

12

Unfold and rotate 180°.

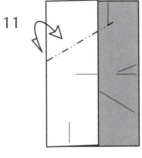

13

Repeat steps 7–12. Turn over and rotate 90°.

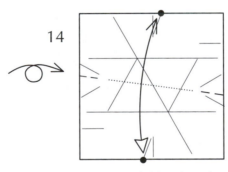

14

Fold and unfold at the edges.

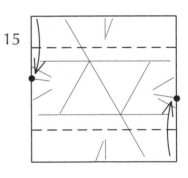

15

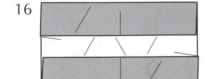

16

17

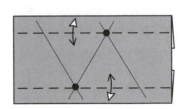

Fold and unfold all the layers together.

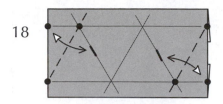

18

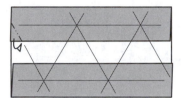

Fold and unfold all
the layers together.

19

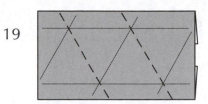

Fold and unfold all the layers
together along the creases.

20

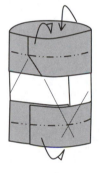

21

Do not crease.

22

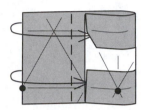

Tuck inside but do not crease.
The dots will meet. Curl the
model to make a cylinder.

23

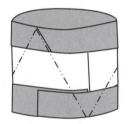

24

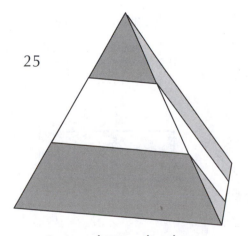

Turn the cylinder
into a tetrahedron.

25

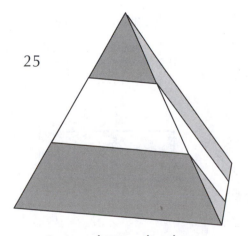

Striped Tetrahedron

Tetrahedron of Triangles

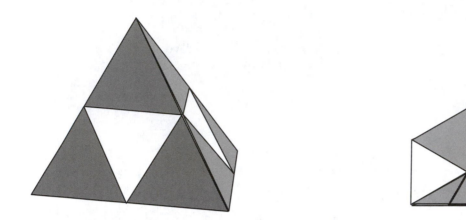

Each face of this tetrahedron has a white inner triangle. The layout shows four triangles in a triangular arrangment, each with inner white triangles. The paper is divided into sevenths.

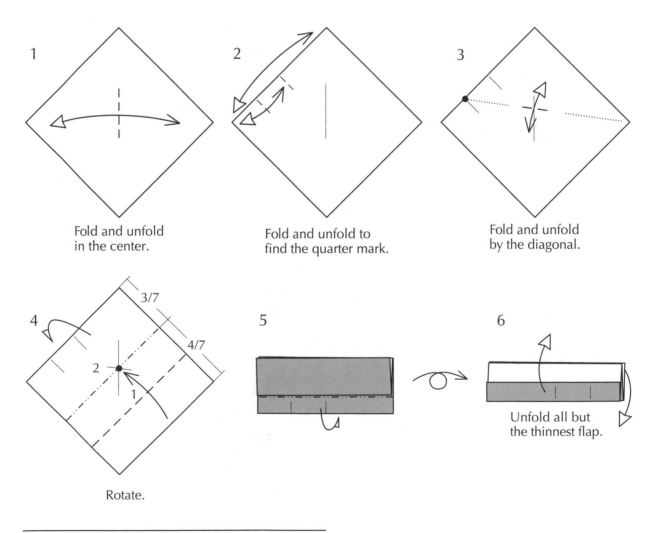

1

Fold and unfold in the center.

2

Fold and unfold to find the quarter mark.

3

Fold and unfold by the diagonal.

4

3/7

4/7

2

1

Rotate.

5

6

Unfold all but the thinnest flap.

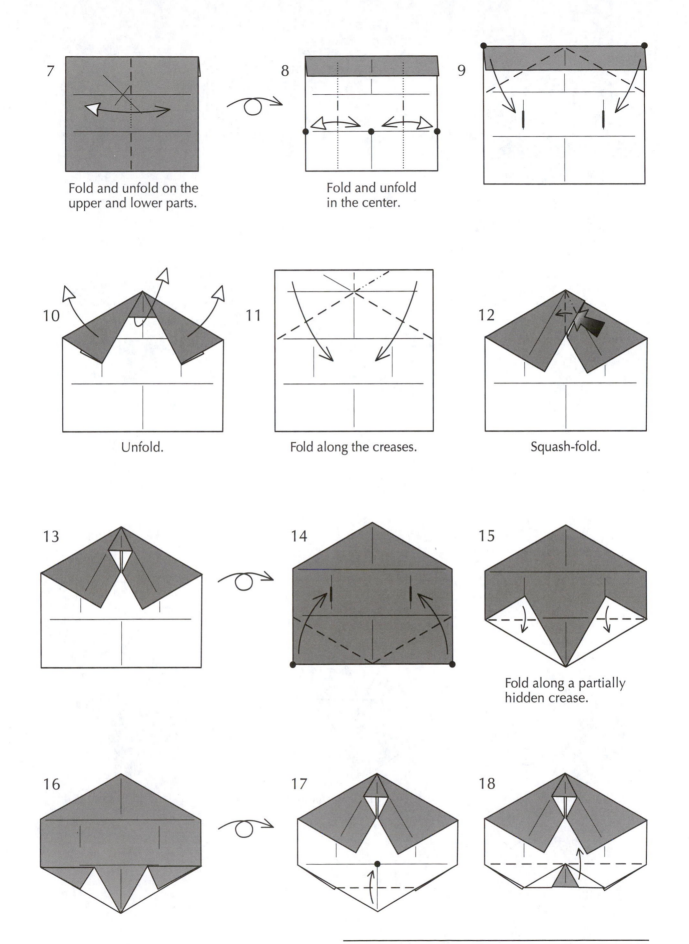

7 Fold and unfold on the
   upper and lower parts.

8 Fold and unfold
   in the center.

9

10 Unfold.

11 Fold along the creases.

12 Squash-fold.

13

14

15 Fold along a partially
    hidden crease.

16

17

18

19

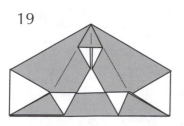

Turn over and rotate 180°.

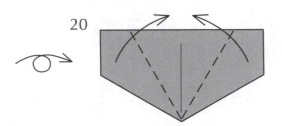

Fold to the center.

21

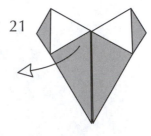

Unfold.

22

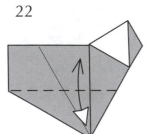

Fold and unfold.

23

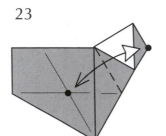

Fold and unfold.

24

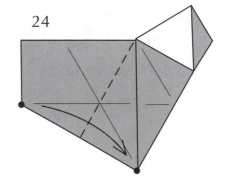

25

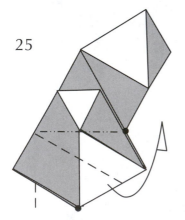

Squash-fold. The dots will meet.

26

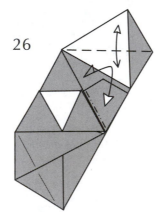

Fold and unfold.

27

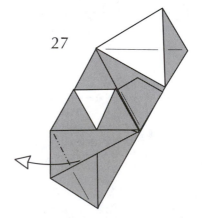

Unfold to step 25.

28

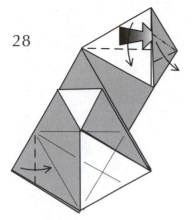

Squash-fold at the top.

29

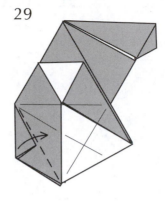

30

A

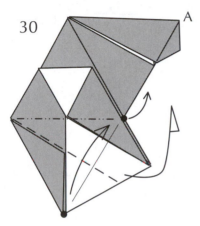

Lift up at the upper dot. Rotate so that *A* is at the bottom.

31

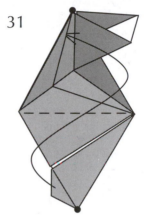

The model is 3D. The view is of the inside of one completed corner of the tetrahedron. Tuck the tab inside. The dots will meet.

32

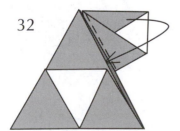

Tuck inside the pocket.

33

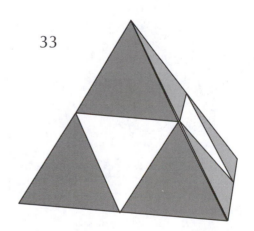

Tetrahedron of Triangles

Stellated Tetrahedron

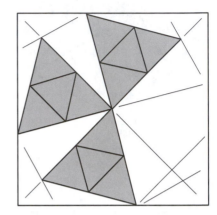

This model resembles four tetrahedra, each on the faces of a central tetrahedron. The crease pattern shows 3/4 square symmetry.

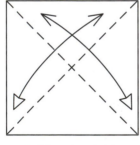

1

Fold and unfold.

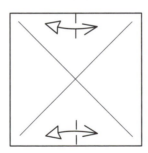

2

Fold and unfold at the top and bottom.

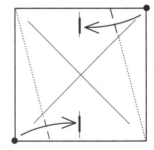

3

Crease at the top and bottom.

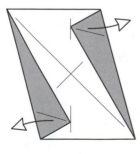

4

Unfold.

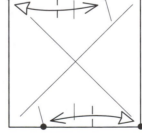

5

Fold and unfold at the top and bottom.

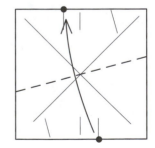

6

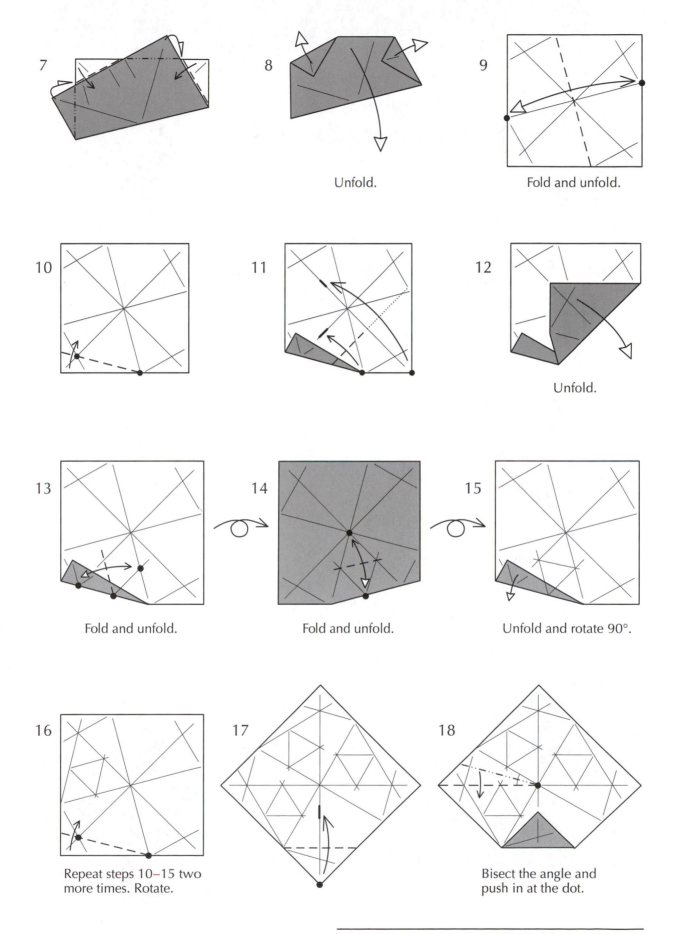

7

8

Unfold.

9

Fold and unfold.

10

11

12

Unfold.

13

Fold and unfold.

14

Fold and unfold.

15

Unfold and rotate 90°.

16

Repeat steps 10–15 two
more times. Rotate.

17

18

Bisect the angle and
push in at the dot.

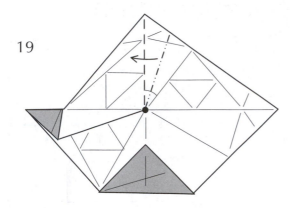

19

Bisect the angle and
push in at the dot.

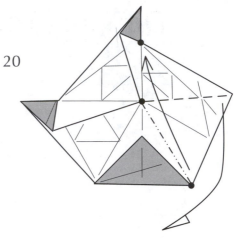

20

Rotate to view the outside so
the center dot is at the top.

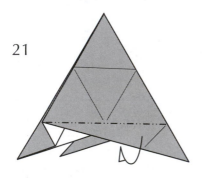

21

Repeat two more times.

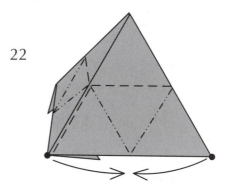

22

Bring the dots together.
Repeat all around. Rotate
to view the bottom.

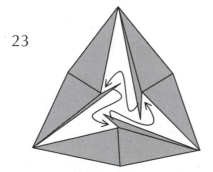

23

Each of the three tabs creates a pocket.
Interlock the tabs into the pockets.

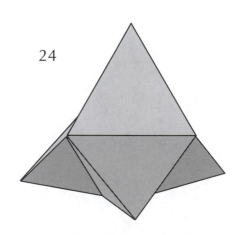

24

Stellated Tetrahedron

Dimpled Truncated Tetrahedron

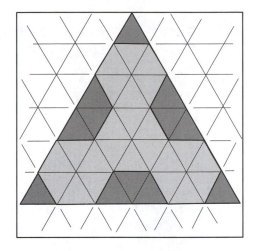

The dimpled truncated tetrahedron is formed from a tetrahedron. In the crease pattern, the darker regions show the sunken parts.

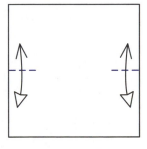

1

Fold and unfold on the left and right.

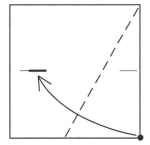

2

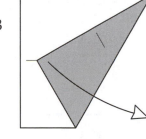

3

Unfold.

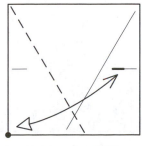

4

Fold and unfold.

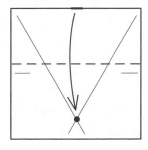

5

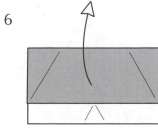

6

Unfold.

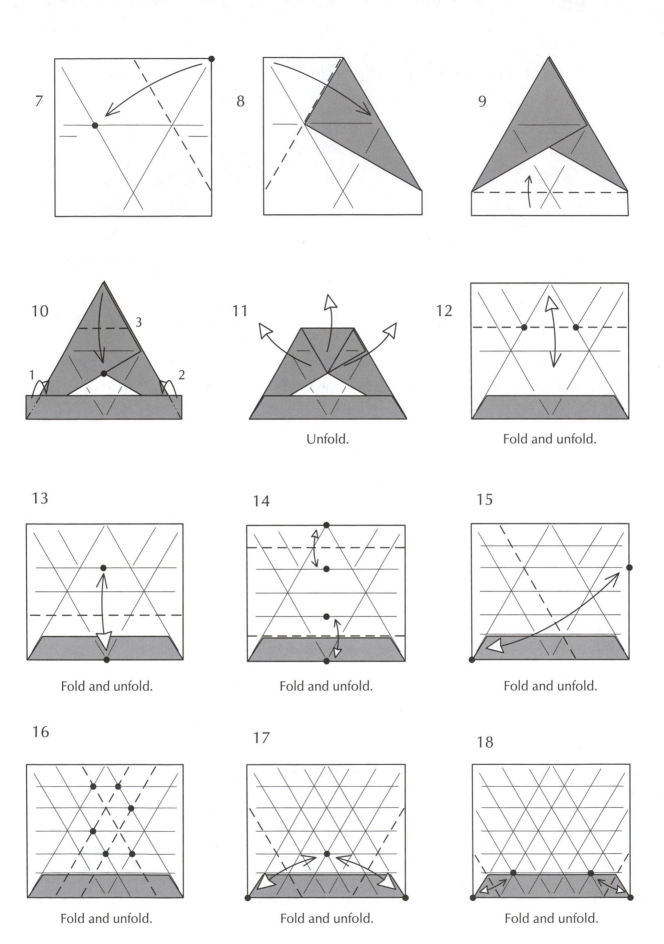

7

8

9

10

11

Unfold.

12

Fold and unfold.

13

Fold and unfold.

14

Fold and unfold.

15

Fold and unfold.

16

Fold and unfold.

17

Fold and unfold.

18

Fold and unfold.

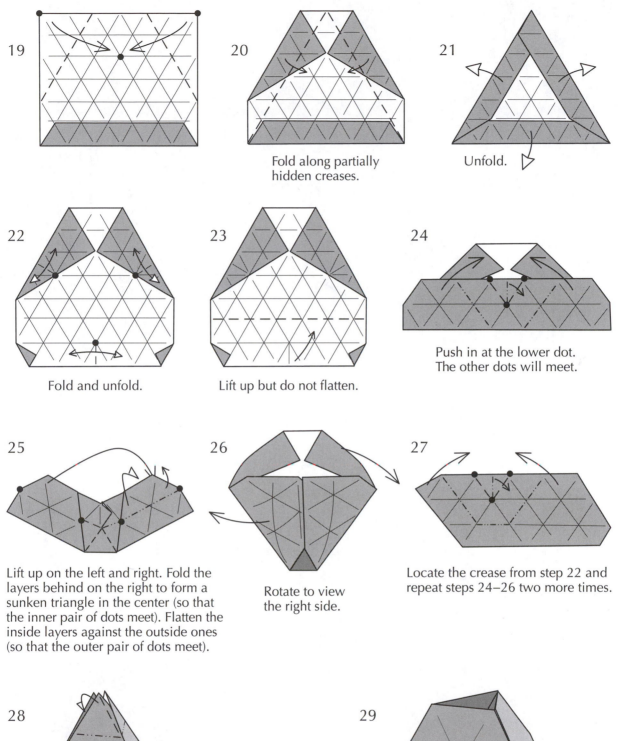

19

20

Fold along partially
hidden creases.

21

Unfold.

22

Fold and unfold.

23

Lift up but do not flatten.

24

Push in at the lower dot.
The other dots will meet.

25

Lift up on the left and right. Fold the
layers behind on the right to form a
sunken triangle in the center (so that
the inner pair of dots meet). Flatten the
inside layers against the outside ones
(so that the outer pair of dots meet).

26

Rotate to view
the right side.

27

Locate the crease from step 22 and
repeat steps 24–26 two more times.

28

Form a sunken triangle with
three interlocking reverse folds.

29

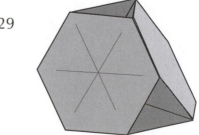

Dimpled Truncated Tetrahedron

Cube Design

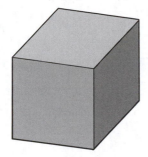

The cube is composed of six square faces. The different layouts of the six faces yield many design possibilities for the cube and related polyhedra.

Layout of the Cube

There are six possible combinations of a band of four with one on each side.

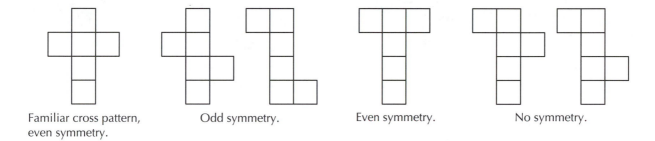

Familiar cross pattern, even symmetry.

Odd symmetry.

Even symmetry.

No symmetry.

There are four arrangements of three squares in a row.

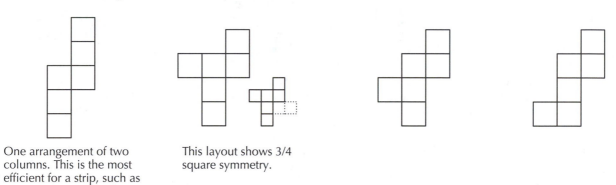

One arrangement of two columns. This is the most efficient for a strip, such as from a dollar bill.

This layout shows 3/4 square symmetry.

There is only one arrangement containing only two squares in a row.

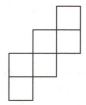

Thus there are eleven nets of the cube.

More Arrangements

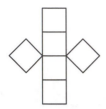

Even/odd symmetry. Sides do not need to meet.

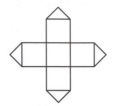

Square symmetry. One face will have lines going through it.

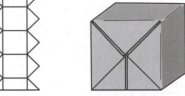

Waterbomb (a traditional origami model).

Crease Patterns

Let s be the length of a side of a small square in a 1×1 square. A larger s value yields a larger cube from the same size paper.

$s = .25$

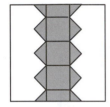 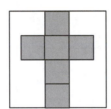 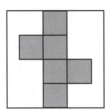 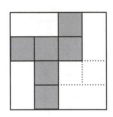

Waterbomb.　　Cross layout.　　Odd symmetry.　　3/4 square symmetry.

$s = \sqrt{2}/5 \approx .283$

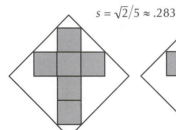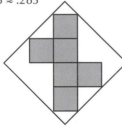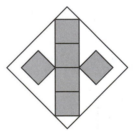

$s = \sqrt{2}/4 \approx .3536$

Diagonal orientation increases efficiency. The first two do not have enough tab to lock, but the third one does, making it my most efficient cube (to be diagrammed in a future volume).

Most efficient, though no tab. Still, can be useful for models using a cube base.

.25 < s < .283　　　$s = 1/(4\cos(\arctan(.5))) \approx .2795$　　　$s = 1/(4\cos(22.5°)) \approx .2706$　　　$s = 1/\sqrt{13} \approx .27735$

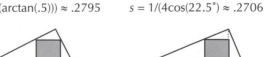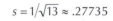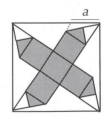

$a = .25$　　　　　$a = 22.5°$

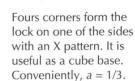

Gives some room for tab.

Most efficient layout for arrangement using 3/4 square symmetry (clean faces). Though not enough tab, useful as a cube base. This crease pattern is used for the Cubehemioctahedron.

Layout for my second most efficient cube, diagrammed here.

Fours corners form the lock on one of the sides with an X pattern. It is useful as a cube base. Conveniently, $a = 1/3$.

Color Patterns

It is interesting to fold a cube with color patterns. Fold the square to show both sides of the paper and then choose a layout to create the pattern. Fold the cube from the layout.

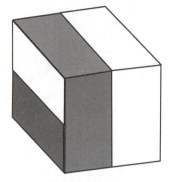

Striped Cube

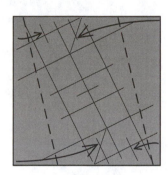

By simply folding two edges as shown, the arrangement of two columns yields a cube with stripes on each face.

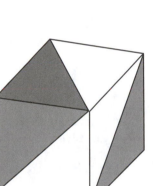

Triangles on Cube

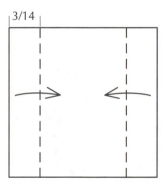

Fold the edges in so that a band goes through the center of the layout.

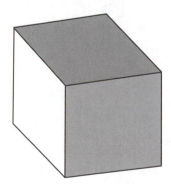

Two opposite faces are white.

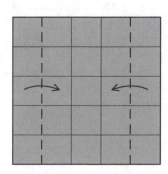

5 × 5 grid.

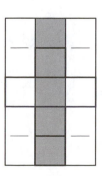

Cross pattern.

Cube Base

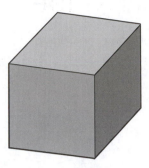

Several polyhedra can be formed using the cube as a base. These models are isomorphic to the cube and share the same surface.

Stellated Octahedron

Cubehemioctahedron

Dimpled Truncated Octahedron

Dimpled Rhombicuboctahedron

Dimpled Great Rhombicuboctahedron

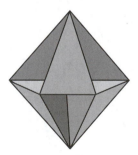

Dimpled Hexagonal Dipyramid

These models are also related to the cube.

Triakis Tetrahedron

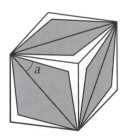

$a = (3/4)\ 45° = 33.75°$

Hide the white paper to form the Triakis Tetrahedron.

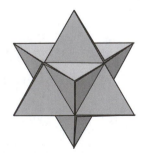

Stella Octangula

This is a union of two tetrahedra.

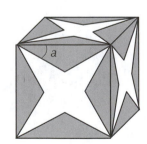

$a = 30°$

Hide the white paper to form the Stella Octangula.

Cube

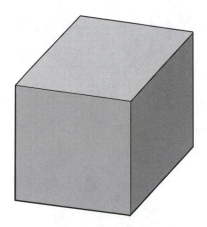

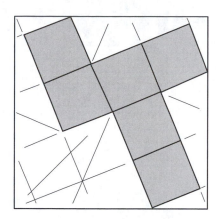

This cube was designed using trapezohedral (or antidiamond) symmetry. The crease pattern shows 3/4 square symmetry.

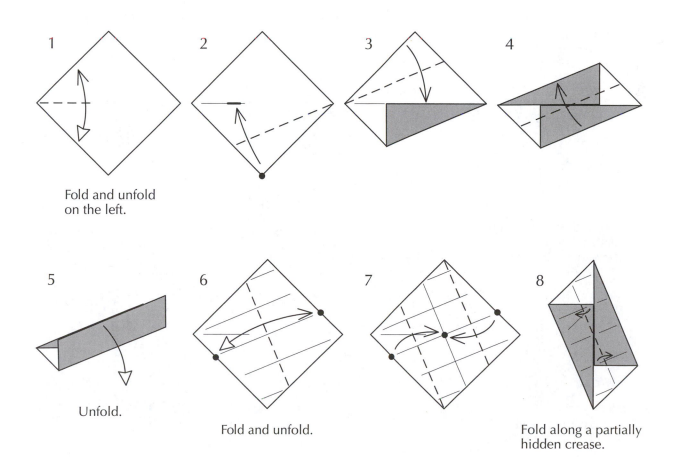

1

Fold and unfold on the left.

2

3

4

5

Unfold.

6

Fold and unfold.

7

8

Fold along a partially hidden crease.

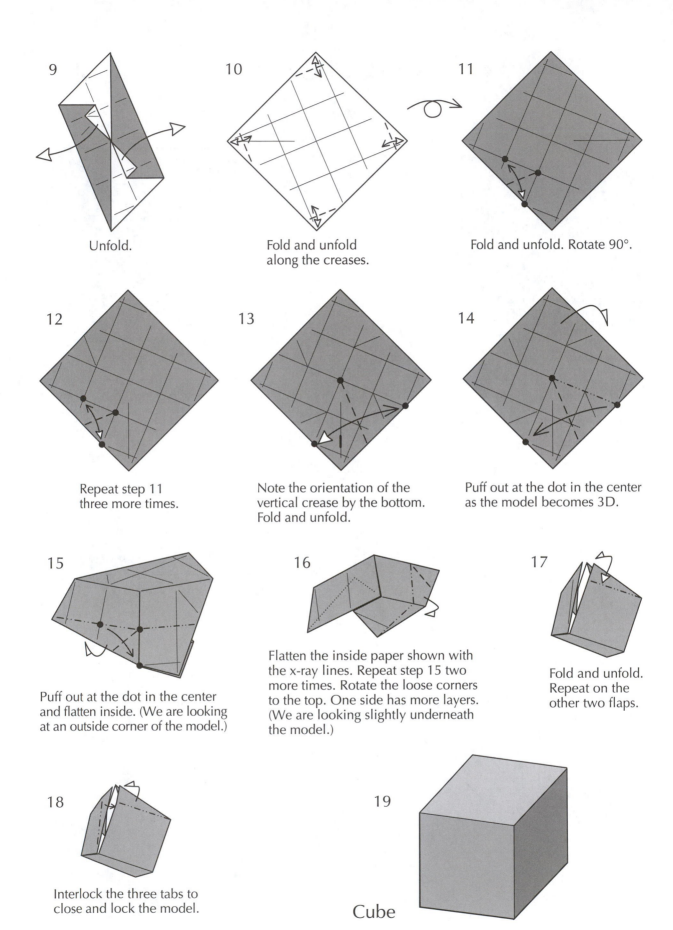

9 Unfold.

10 Fold and unfold along the creases.

11 Fold and unfold. Rotate 90°.

12 Repeat step 11 three more times.

13 Note the orientation of the vertical crease by the bottom. Fold and unfold.

14 Puff out at the dot in the center as the model becomes 3D.

15 Puff out at the dot in the center and flatten inside. (We are looking at an outside corner of the model.)

16 Flatten the inside paper shown with the x-ray lines. Repeat step 15 two more times. Rotate the loose corners to the top. One side has more layers. (We are looking slightly underneath the model.)

17 Fold and unfold. Repeat on the other two flaps.

18 Interlock the three tabs to close and lock the model.

19 Cube

Striped Cube

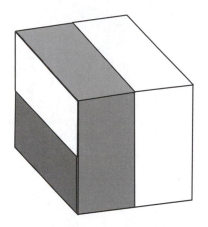

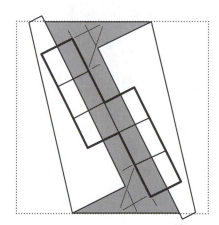

As seen with the tetrahedra designs, color patterns make for effective models; we now do the same with cube designs. The layout used here is two bands of three squares. By folding two edges toward the center, a layout with stripes is formed, as shown above.

1

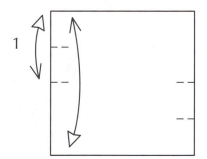

Make small marks by folding and unfolding in quarters.

2

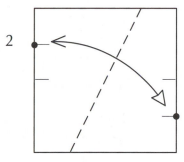

Fold and unfold.

3

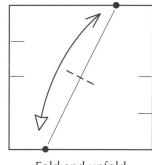

Fold and unfold in the center.

4

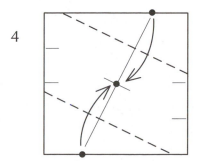

5

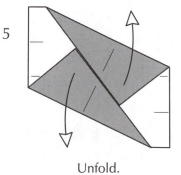

Unfold.

6

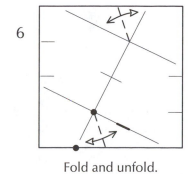

Fold and unfold.

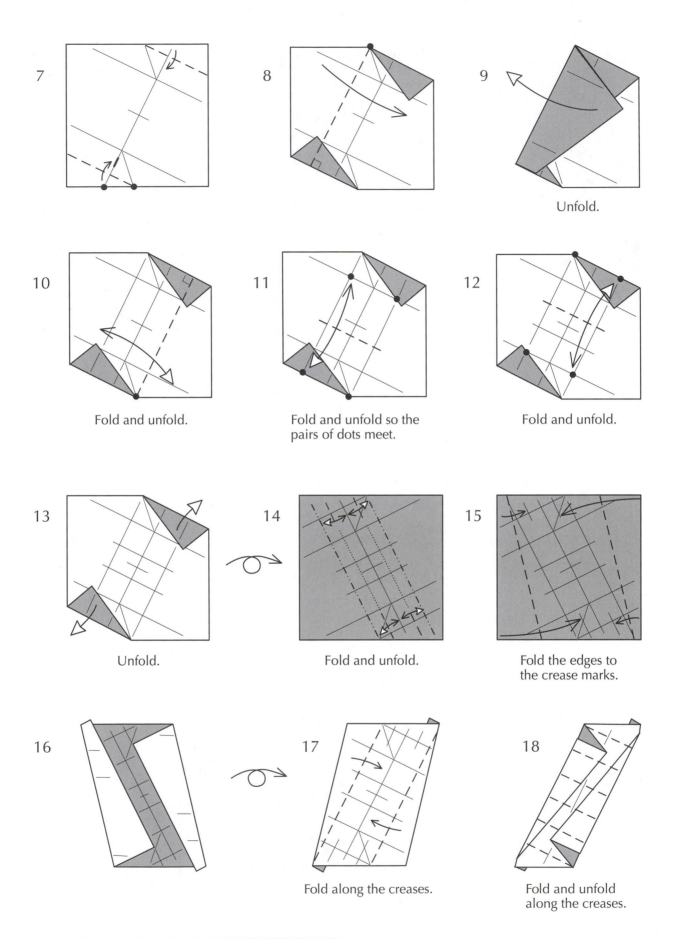

7

8

9

Unfold.

10

Fold and unfold.

11

Fold and unfold so the
pairs of dots meet.

12

Fold and unfold.

13

Unfold.

14

Fold and unfold.

15

Fold the edges to
the crease marks.

16

17

Fold along the creases.

18

Fold and unfold
along the creases.

19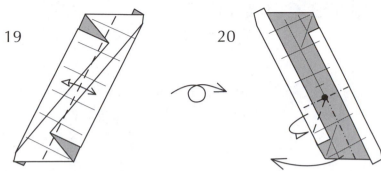

Fold and unfold.

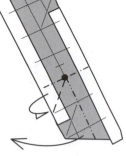

20

Puff out at the dot.

21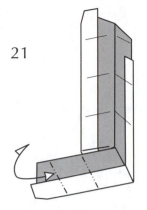

Fold and unfold along
the creases. Rotate.

22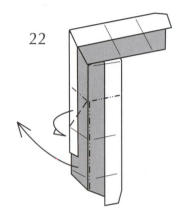

Repeat steps 20–21.

23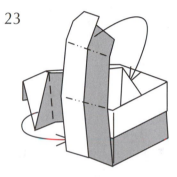

Tuck inside.

24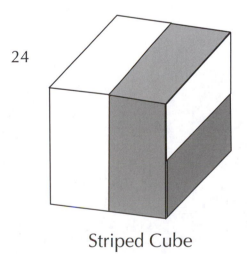

Striped Cube

Triangles on Cube

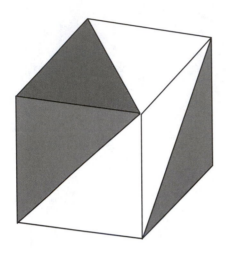

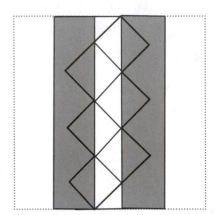

The layout shows three rows of two squares. The paper is divided into sevenths.

1

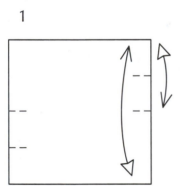

Make small marks by folding and unfolding in quarters.

2

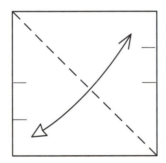

Fold and unfold.

3

Fold and unfold creasing along the diagonal.

4

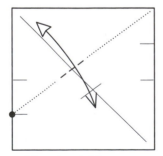

Fold and unfold creasing along the diagonal.

5

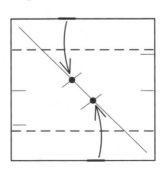

6

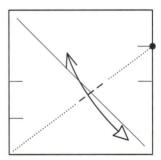

Fold along the creases.

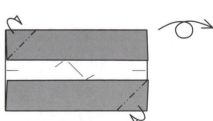

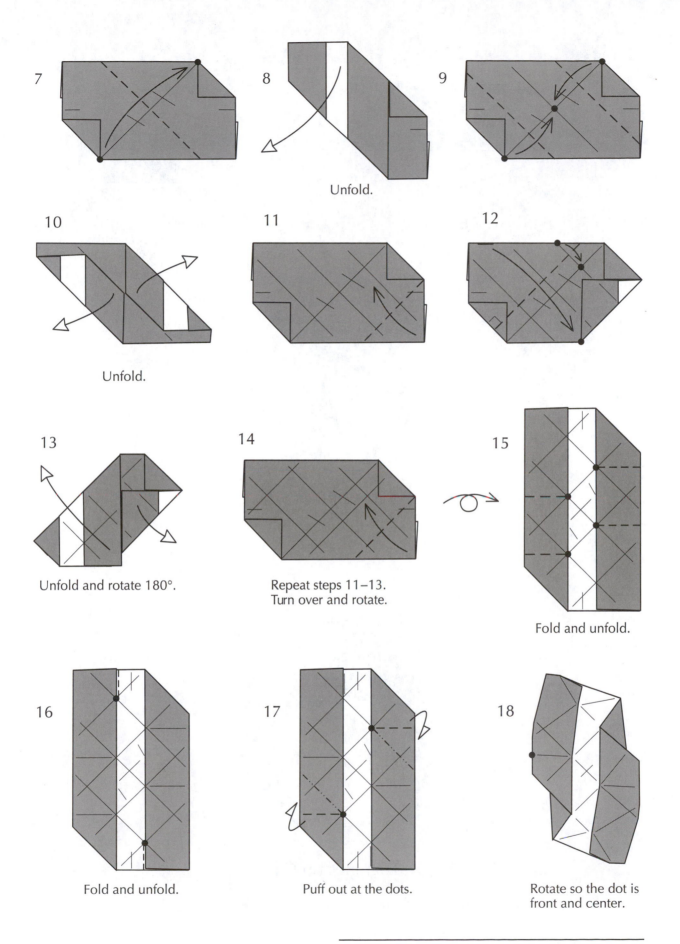

7

8

Unfold.

9

10

Unfold.

11

12

13

Unfold and rotate 180°.

14

Repeat steps 11–13.
Turn over and rotate.

15

Fold and unfold.

16

Fold and unfold.

17

Puff out at the dots.

18

Rotate so the dot is
front and center.

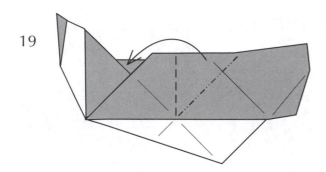

19

Turn over and repeat.

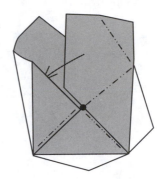

20

Puff out at the dot. Begin to tuck inside. Turn over and repeat. Rotate to view the opening.

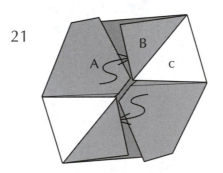

21

A   B   C

Region A goes under B and tucks into the pocket under C. Repeat below simultaneously.

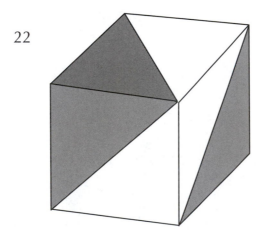

22

Triangles on Cube

Cube with Squares

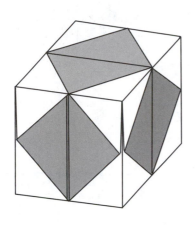

Each face of this cube has a diamond color pattern. The paper is divided into ninths.

1

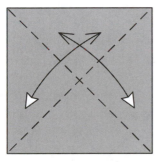

Fold and unfold
along the diagonals.

2

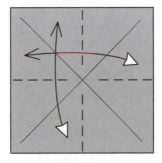

Fold and unfold but
not in the center.

3

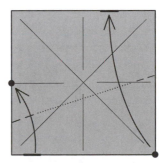

Bring the lower right corner to
the top edge and the bottom
edge to the left center. Crease
on the left and right.

4

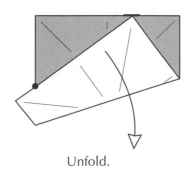

Unfold.

5

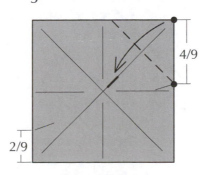

4/9

2/9

The 2/9 and 4/9
marks are found.

6

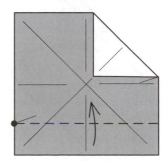

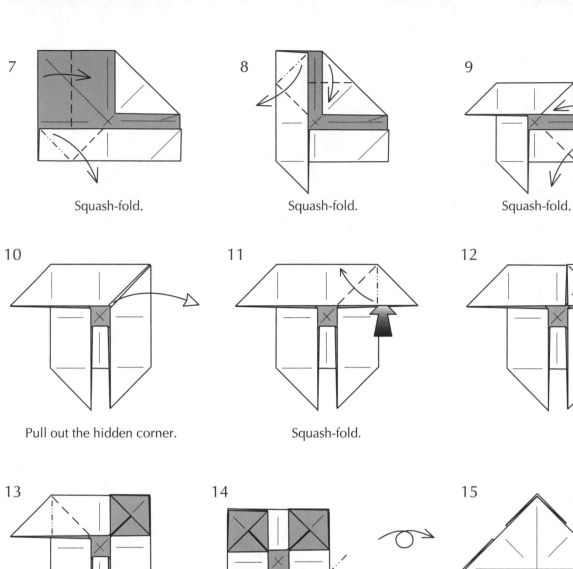

7

Squash-fold.

8

Squash-fold.

9

Squash-fold.

10

Pull out the hidden corner.

11

Squash-fold.

12

13

Repeat steps 11–12 on
the three other corners.

14

Turn over and rotate.

15

16

Repeat steps 14–15
three more times.

17

Fold and unfold. Rotate 90°.

18

Repeat step 17
three more times.

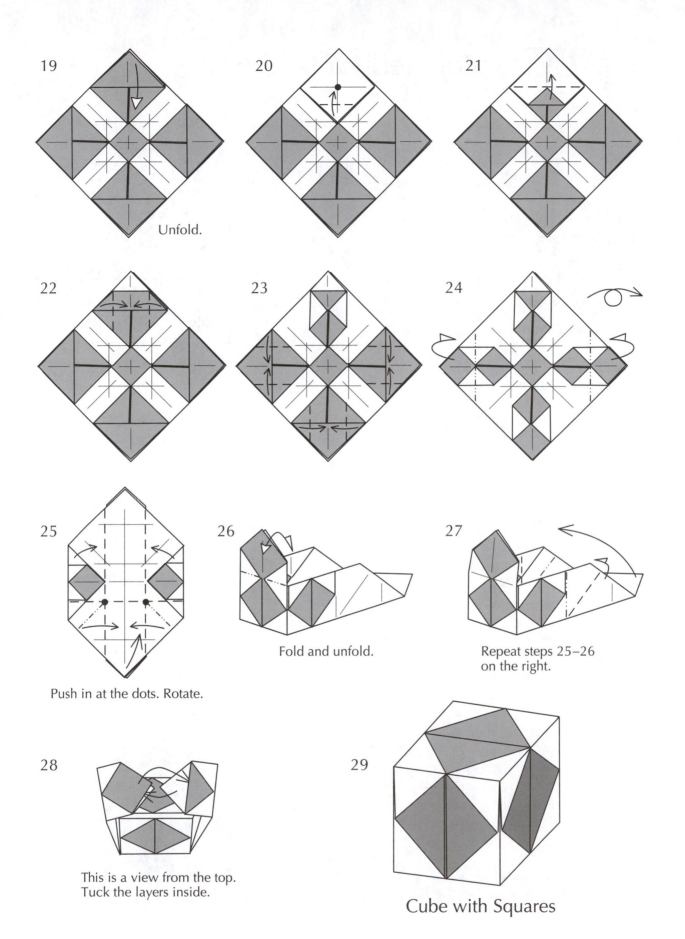

19

20

21

Unfold.

22

23

24

25

Push in at the dots. Rotate.

26

Fold and unfold.

27

Repeat steps 25–26
on the right.

28

This is a view from the top.
Tuck the layers inside.

29

Cube with Squares

Stellated Octahedron

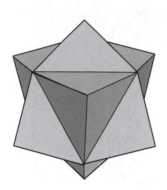

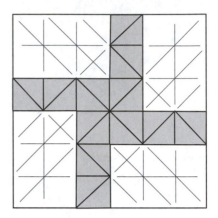

The stellated octahedron has the same surface as a cube. The layout shows square symmetry. The paper is divided into sixths.

1

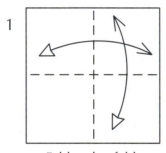

Fold and unfold.

2

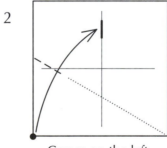

Crease on the left.

3

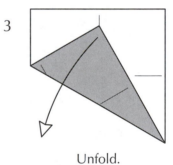

Unfold.

4

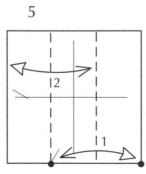

Fold and unfold
at the bottom.

5

Fold and unfold.

6

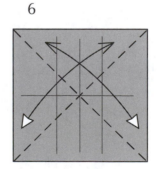

Fold and unfold.

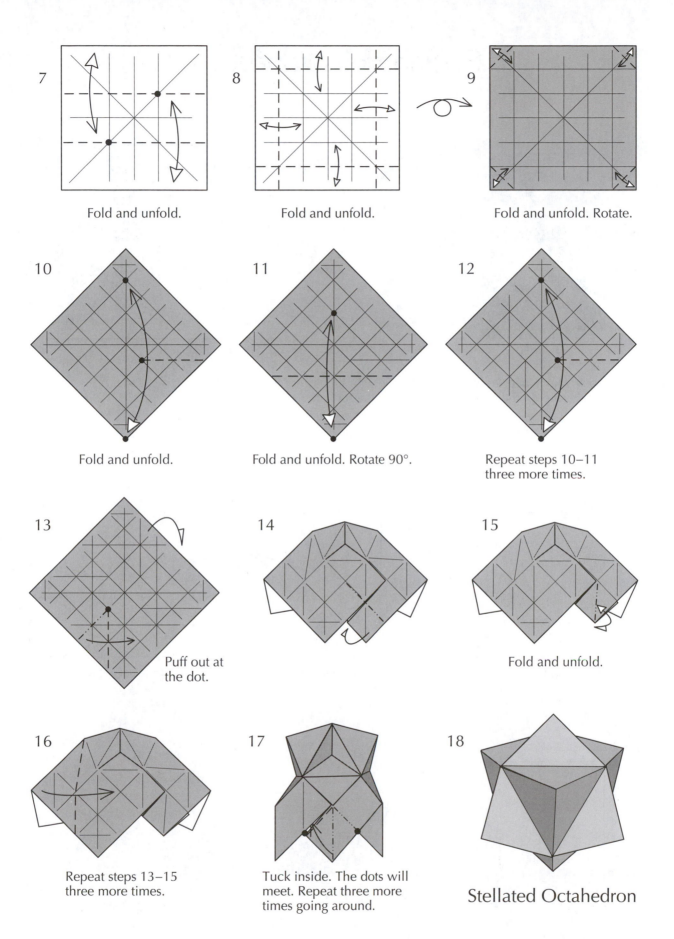

7 Fold and unfold.

8 Fold and unfold.

9 Fold and unfold. Rotate.

10 Fold and unfold.

11 Fold and unfold. Rotate 90°.

12 Repeat steps 10–11 three more times.

13 Puff out at the dot.

14

15 Fold and unfold.

16 Repeat steps 13–15 three more times.

17 Tuck inside. The dots will meet. Repeat three more times going around.

18 Stellated Octahedron

Cubehemioctahedron

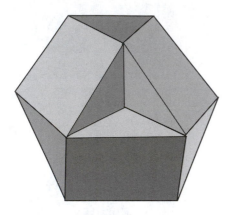

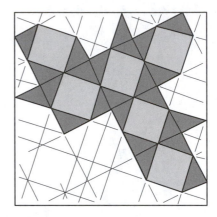

The cubehemioctahedron is basically a cube with sunken corners. This model uses 3/4 square symmetry. The darker paper in the crease pattern shows the sunken sides.

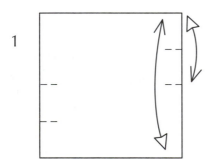

1

Make small marks by folding and unfolding in quarters.

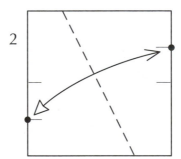

2

Fold and unfold.

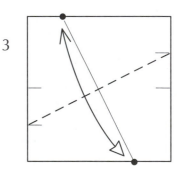

3

Fold and unfold.

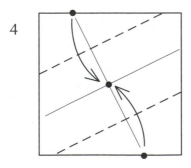

4

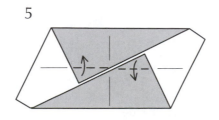

5

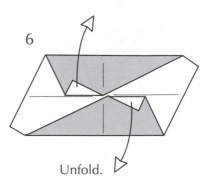

6

Unfold.

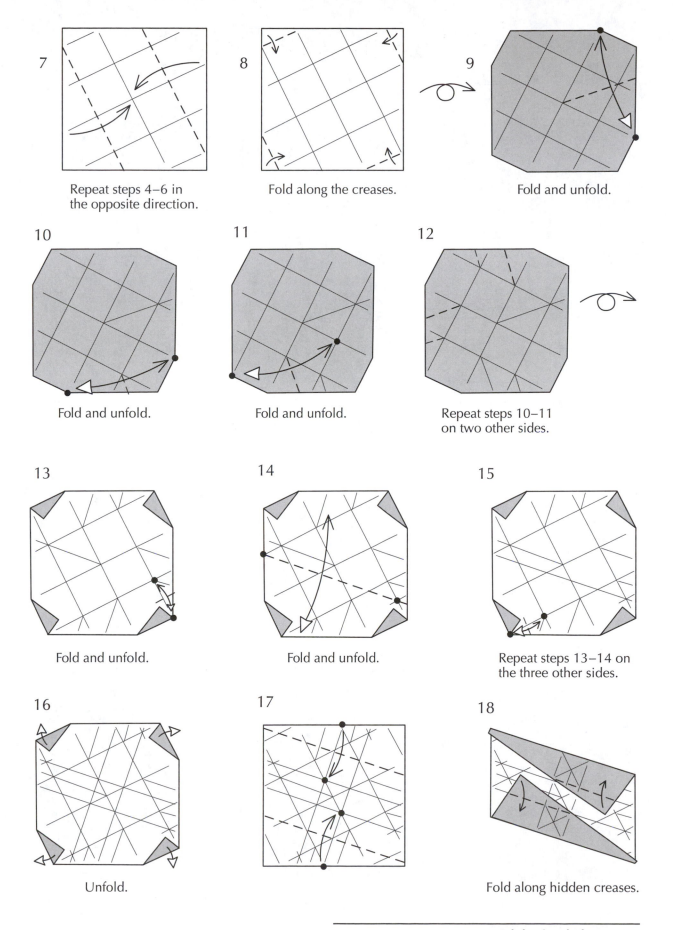

7 Repeat steps 4–6 in the opposite direction.

8 Fold along the creases.

9 Fold and unfold.

10 Fold and unfold.

11 Fold and unfold.

12 Repeat steps 10–11 on two other sides.

13 Fold and unfold.

14 Fold and unfold.

15 Repeat steps 13–14 on the three other sides.

16 Unfold.

17

18 Fold along hidden creases.

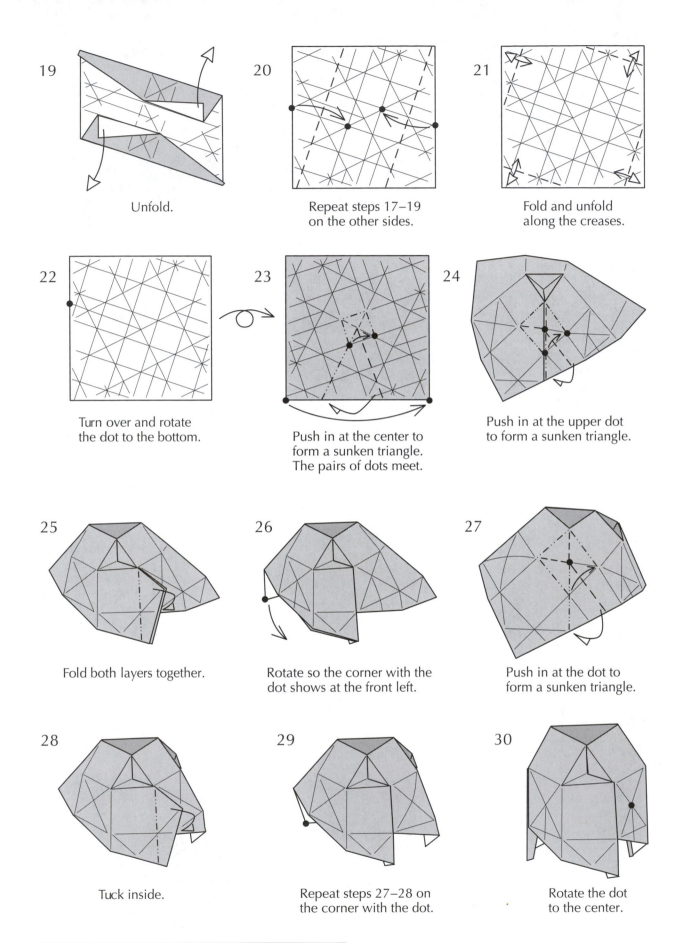

19 Unfold.

20 Repeat steps 17–19 on the other sides.

21 Fold and unfold along the creases.

22 Turn over and rotate the dot to the bottom.

23 Push in at the center to form a sunken triangle. The pairs of dots meet.

24 Push in at the upper dot to form a sunken triangle.

25 Fold both layers together.

26 Rotate so the corner with the dot shows at the front left.

27 Push in at the dot to form a sunken triangle.

28 Tuck inside.

29 Repeat steps 27–28 on the corner with the dot.

30 Rotate the dot to the center.

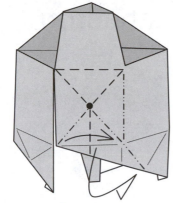

31

Push in at the dot to
form a sunken triangle.

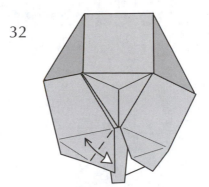

32

Fold and unfold
along the crease.

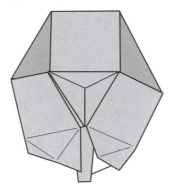

33

Repeat steps 31–32 two
more times going around.

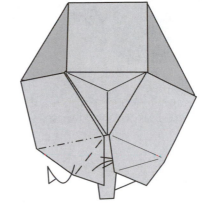

34

Begin to form the bottom sunken
triangle by folding toward the
inside center and tucking.

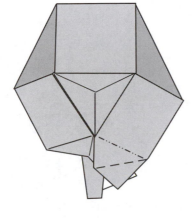

35

Repeat step 34 two more
times to complete the
bottom sunken triangle.

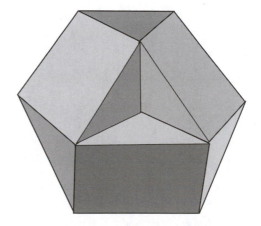

36

Cubehemioctahedron

Dimpled Rhombicuboctahedron

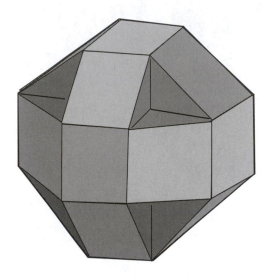

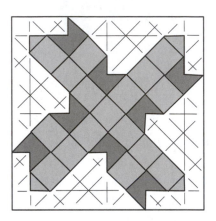

The dimpled rhombicuboctahedron has the same surface as the cube. The dark regions of the crease pattern show the sunken sides. This model uses square symmetry. The paper is divided into tenths.

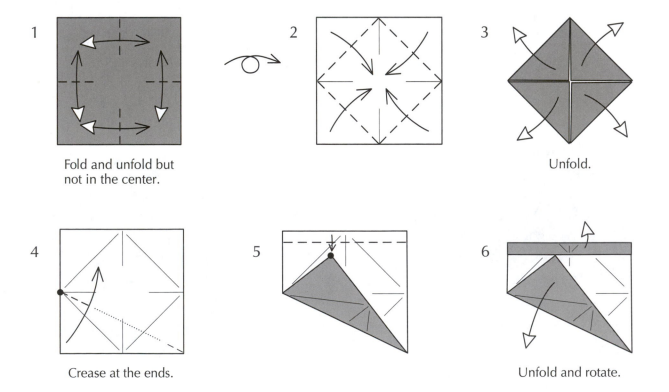

1   Fold and unfold but not in the center.

2

3   Unfold.

4   Crease at the ends.

5

6   Unfold and rotate.

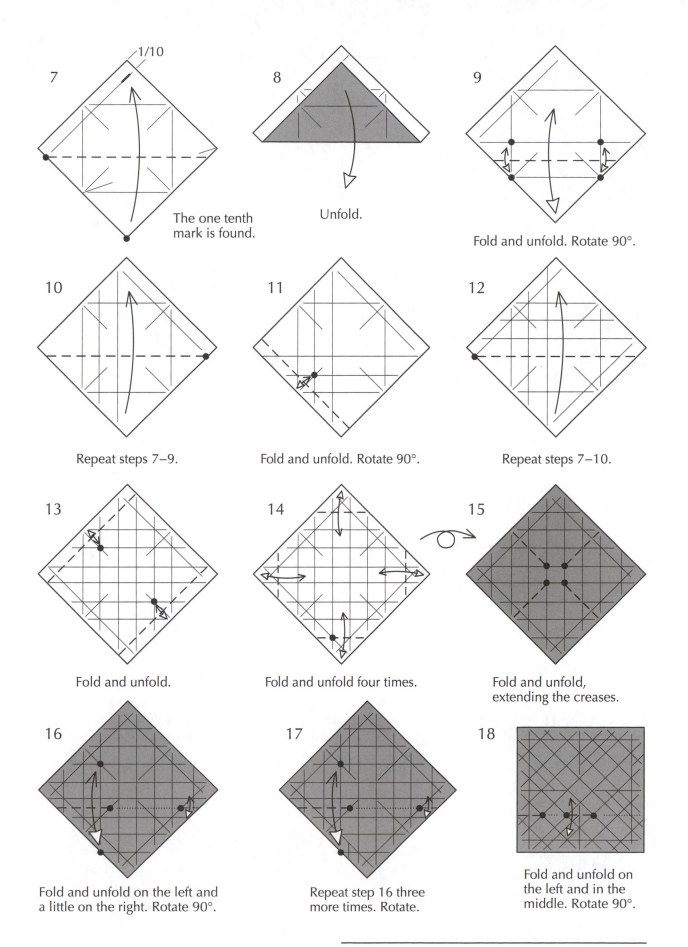

7

1/10

The one tenth
mark is found.

8

Unfold.

9

Fold and unfold. Rotate 90°.

10

Repeat steps 7–9.

11

Fold and unfold. Rotate 90°.

12

Repeat steps 7–10.

13

Fold and unfold.

14

Fold and unfold four times.

15

Fold and unfold,
extending the creases.

16

Fold and unfold on the left and
a little on the right. Rotate 90°.

17

Repeat step 16 three
more times. Rotate.

18

Fold and unfold on
the left and in the
middle. Rotate 90°.

19

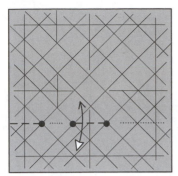

Repeat step 18 three more times.

20

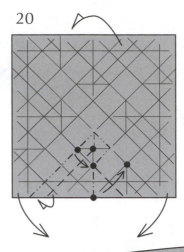

Push in at the upper center dot to form a sunken triangle. The other pairs of dots meet.

21

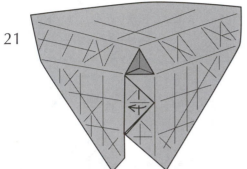

22

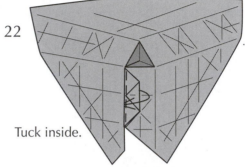

Tuck inside.

23

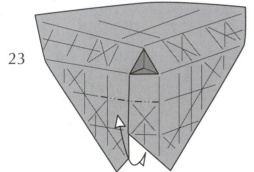

Fold and unfold all the layers.

24

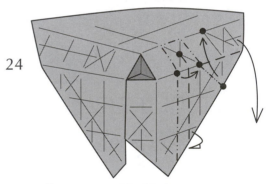

Repeat steps 20–23 three more times. Rotate the top to the bottom.

25

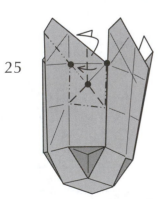

Push in at the lower dot. Rotate a little to the right.

26

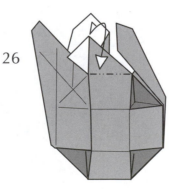

Fold and unfold.

27

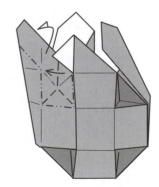

Repeat steps 25–26 three more times.

28

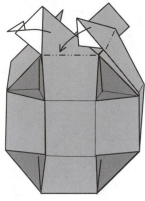

Mountain-fold to form
a flat square with four
white triangles.

29

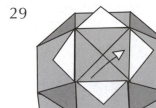

Unfold back to step 28.

30

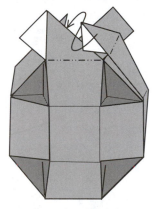

Tuck each tab to form
a four-way twist lock.

31

Rotate the top
to the bottom.

32

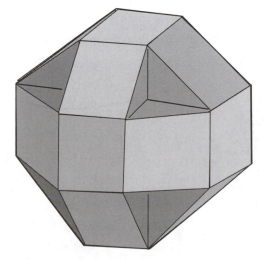

Dimpled Rhombicuboctahedron

Stacked Cubes

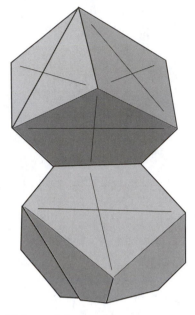

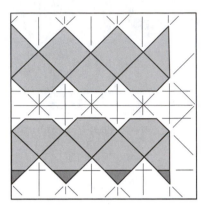

This model resembles two connected cubes. The paper is divided into sevenths. The crease pattern shows that the model uses even symmetry. The darker regions refer to the sunken triangle at the bottom, on which the model stands.

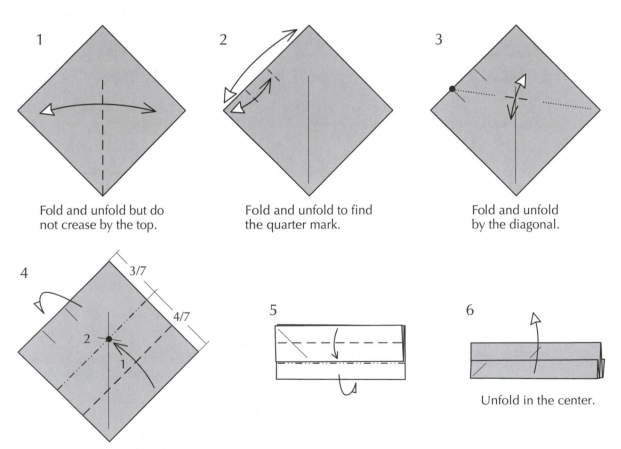

1

Fold and unfold but do not crease by the top.

2

Fold and unfold to find the quarter mark.

3

Fold and unfold by the diagonal.

4

3/7

4/7

2

1

Rotate.

5

6

Unfold in the center.

7

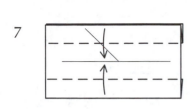

8

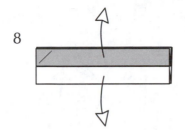

Unfold and rotate.

9

Fold and unfold.

10

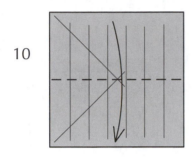

11

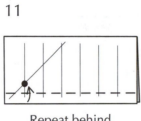

Repeat behind.

12

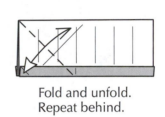

Fold and unfold.
Repeat behind.

13

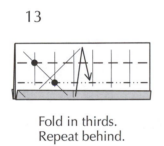

Fold in thirds.
Repeat behind.

14

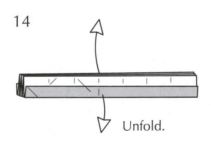

Unfold.

15

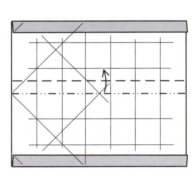

Mountain-fold along the crease.

16

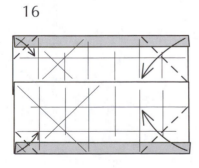

17

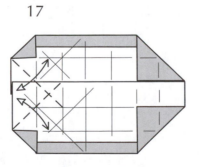

Fold and unfold.

18

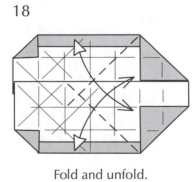

Fold and unfold.

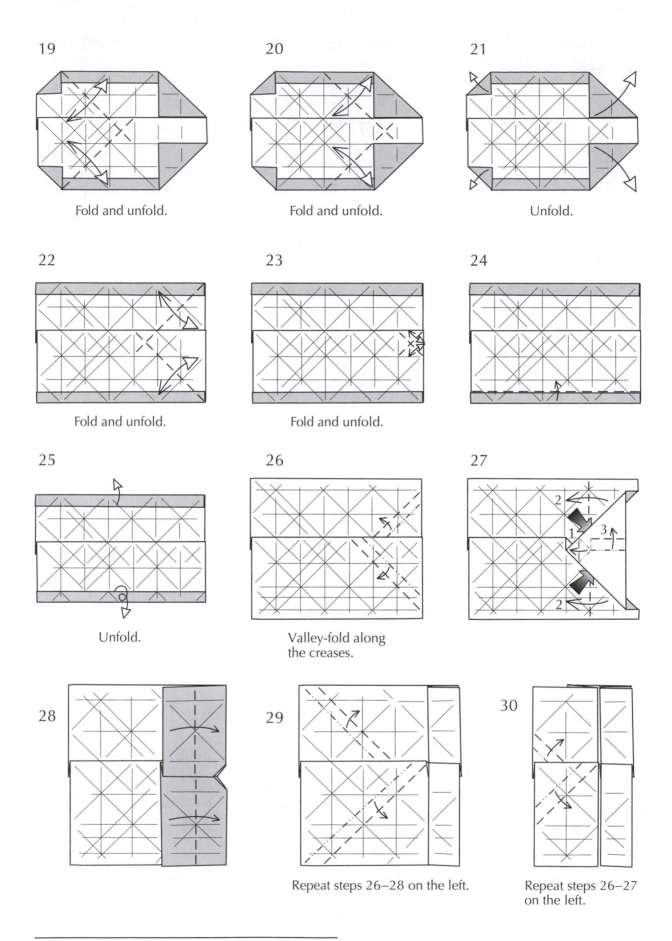

19

Fold and unfold.

20

Fold and unfold.

21

Unfold.

22

Fold and unfold.

23

Fold and unfold.

24

25

Unfold.

26

Valley-fold along
the creases.

27

28

29

Repeat steps 26–28 on the left.

30

Repeat steps 26–27
on the left.

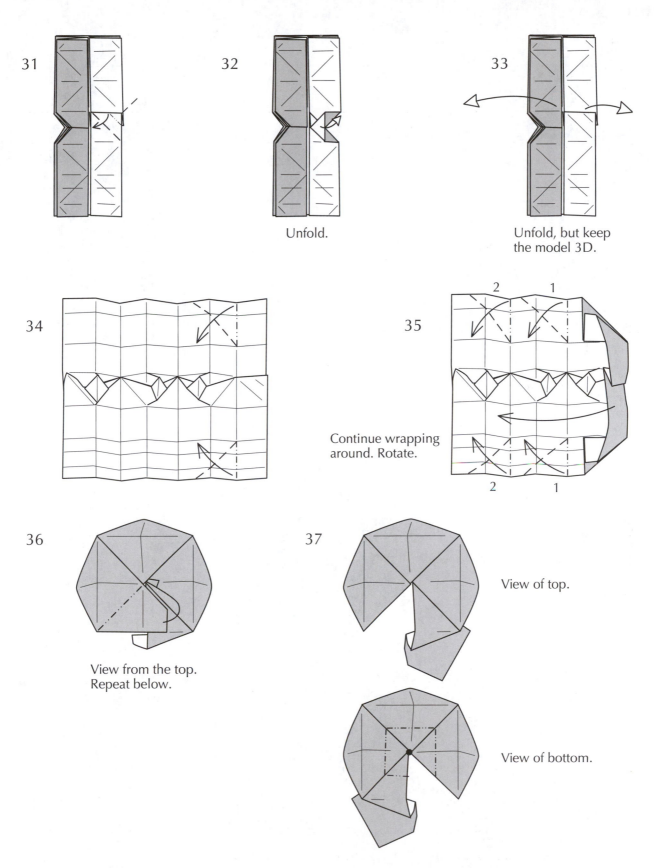

31

32

Unfold.

33

Unfold, but keep the model 3D.

34

35

Continue wrapping around. Rotate.

36

View from the top. Repeat below.

37

View of top.

View of bottom.

Push in on the bottom. This will form a sunken triangle, on which the model will stand.

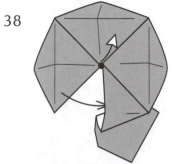

38

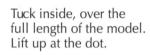

View of top.

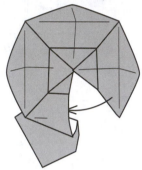

View of bottom.

Tuck inside, over the
full length of the model.
Lift up at the dot.

39

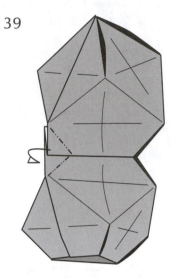

Open some layers to connect
and lock while refolding the
edge from step 31.

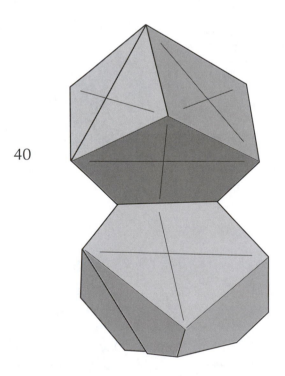

40

Stacked Cubes

Octahedron Design

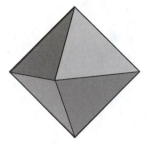

If $b = 1$ then
$h = \sqrt{3}/2$

The octahedron is composed of eight equilateral triangles. It can be used as a base to create more polyhedra.

Layout of the Octahedron

Here is a selection of some of the possible layouts.

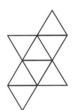  

Band of six triangles.

  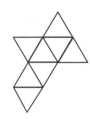

Odd symmetry.

No symmetry.

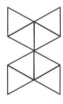 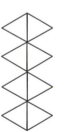

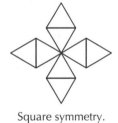

Even/odd symmetry.

Odd symmetry.

Square symmetry.

Crease Patterns

Let s be the length of a side of a triangle in a 1×1 square. A larger s value yields a larger octahedron from the same size paper.

$s = \sqrt{2}/(3 + \sqrt{3}) \approx .299$

$s = 1/3 \approx .33333$

$s = 1/(2\sqrt{3}) \approx .2887$

$s = 2/(3\sqrt{3}) \approx .3849$

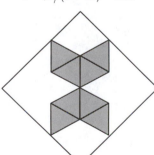

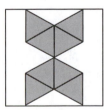

Not enough tab.

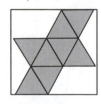

Not enough tab.

Even/odd symmetry.

Odd symmetry.

$s = \sqrt{3}/5 \approx .34641$

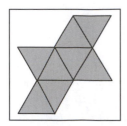

$s \approx .3239$

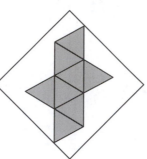

$s = 1/\sqrt{6} \approx .408$

$s = \sqrt{123}/30 \approx .3697$

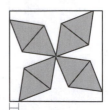

1/10

To be diagrammed in "Antiprism Design" in a future volume.

Odd symmetry, band along diagonal.

Square symmetry yields the most efficient layout but not enough tab. Still, it can be useful as an octahedral base.

My most efficient design. By rotating the diagonal a small amount, the four corners can lock the model. Diagrammed.

Color Patterns

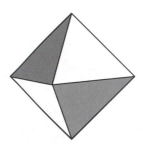

Duo-Colored Octahedron

The triangles shown by the dotted lines will soon be white.

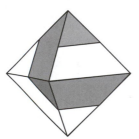

Striped Octahedron

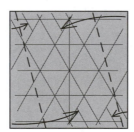

By folding two edges as shown, the arrangement of two columns yields an octahedron with stripes on each face.

Octahedral Base

Several polyhedra can be formed using the octahedron as a base.

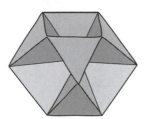

Octahemioctahedron

Dimpled Truncated Octahedron

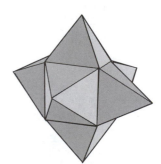

Stellated Cube

These three models have the same surface as the octahedron. The octahemioctahedron and dimpled truncated octahedron are formed by sinking the six vertices.

114   Part II: Platonic and Related Polyhedra

Octahedron

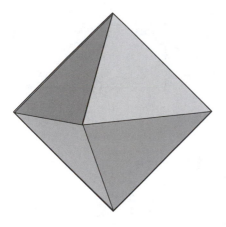

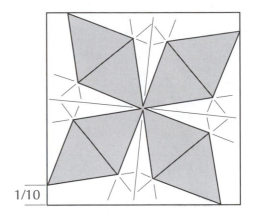

1/10

The design for this octahedron has square symmetry. The model closes with four thin tabs interlocking at the top, a concept called a twist lock. The thin tabs allow for efficient use of paper.

1

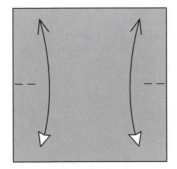

Fold and unfold on the left and right.

2

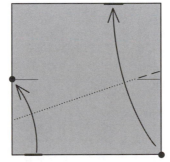

Bring the lower right corner to the top edge and the bottom edge to the left center. Crease on the right.

3

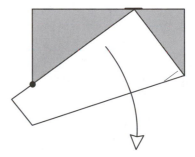

Unfold and rotate 180°.

4

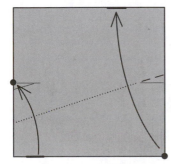

Repeat steps 2–3 and rotate 90°.

5

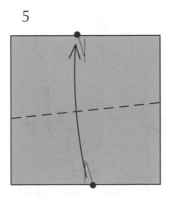

6

Fold and unfold. Turn over and repeat.

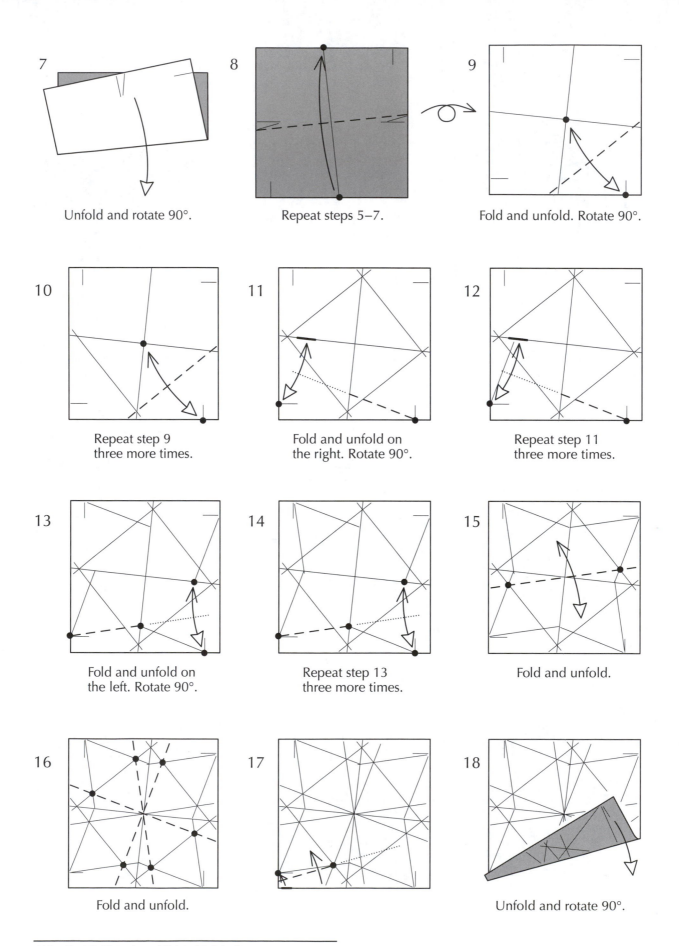

7

Unfold and rotate 90°.

8

Repeat steps 5–7.

9

Fold and unfold. Rotate 90°.

10

Repeat step 9
three more times.

11

Fold and unfold on
the right. Rotate 90°.

12

Repeat step 11
three more times.

13

Fold and unfold on
the left. Rotate 90°.

14

Repeat step 13
three more times.

15

Fold and unfold.

16

Fold and unfold.

17

18

Unfold and rotate 90°.

19

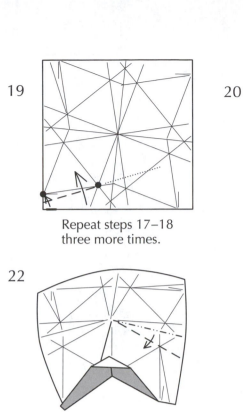

Repeat steps 17–18
three more times.

20

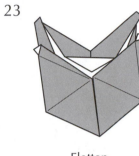

Push in at the dot.

21

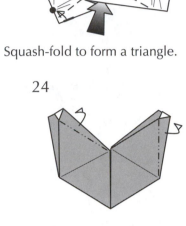

Squash-fold to form a triangle.

22

The angles of the white triangle are not
important. Repeat steps 20–21 three
more times. Rotate to view the outside.

23

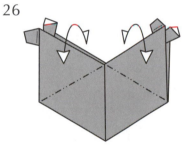

Flatten.

24

Turn over and repeat.

25

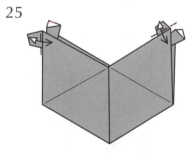

Fold and unfold.
Repeat behind.

26

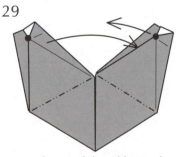

Open to fold inside and
unfold. Repeat behind.

27

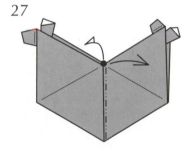

Open and flatten. Follow
the dot in the next step.

28

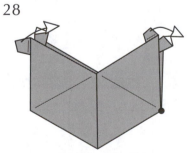

Unfold the thin flaps.
Repeat behind.

29

Open the model and bring the
dots together. Close the model
by interlocking the four tabs. The
tabs spiral inward. This method
is called a twist lock. (Fold along
the creases in step 25.)

30

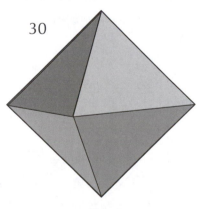

Octahedron

Striped Octahedron

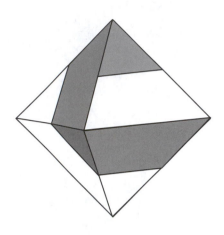

Each face of this octahedron is striped, and this model is similar to the Striped Cube. The layout shows two bands of four triangles each.

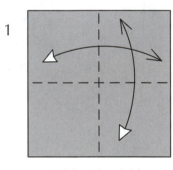 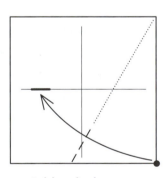

1 Fold and unfold.

2 Fold at the bottom.

3 Unfold and rotate 180°.

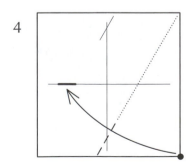 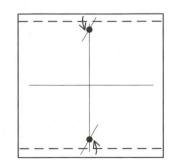 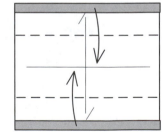

4 Repeat steps 2–3.

5

6

7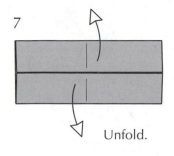

Unfold.

8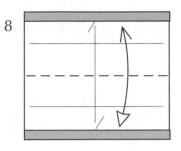

Fold and unfold.

9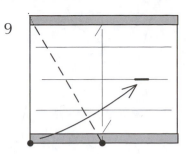

10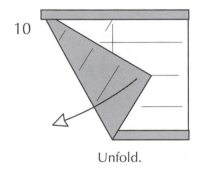

Unfold.

11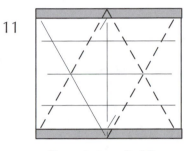

Repeat steps 9–10
three more times.

12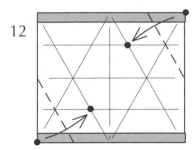

13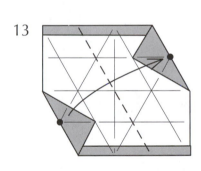

14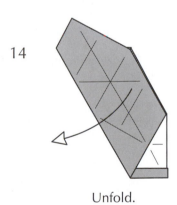

Unfold.

15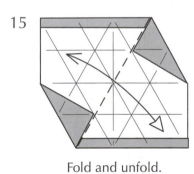

Fold and unfold.

16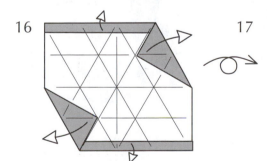

Unfold.

17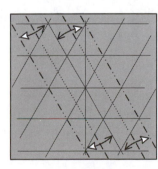

Fold and unfold.

18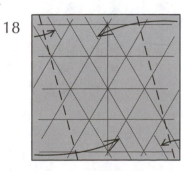

Fold the edges to
the crease marks.

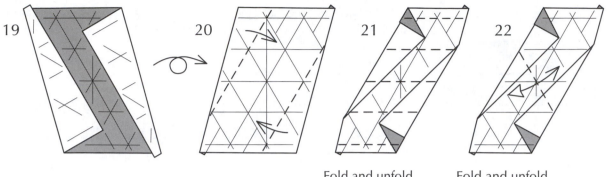

19

20

21

Fold and unfold.

22

Fold and unfold.

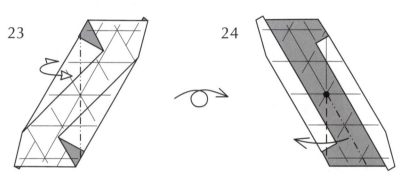

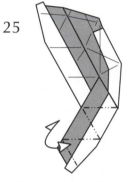

23

Fold and unfold.

24

Puff out at the dot.

25

Fold and unfold along
the creases. Rotate.

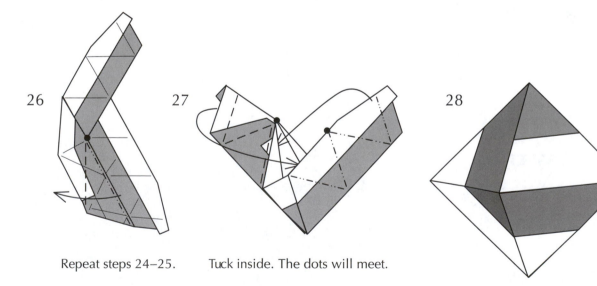

26

Repeat steps 24–25.

27

Tuck inside. The dots will meet.

28

Striped Octahedron

Duo-Colored Octahedron

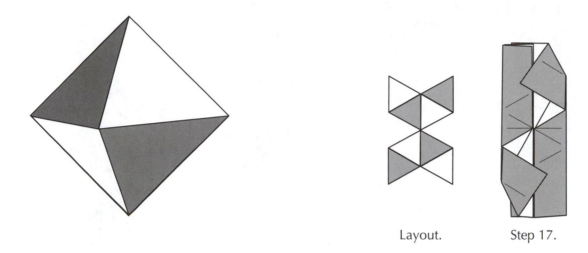

Layout.   Step 17.

Four sides of this octahedron are of one color, and the other four are white. Step 17 shows the formation of two white triangles. By step 27 all four white triangles are formed.

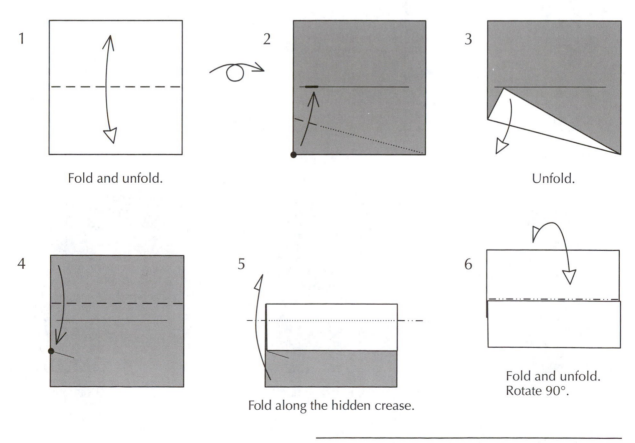

1 Fold and unfold.

2

3 Unfold.

4

5 Fold along the hidden crease.

6 Fold and unfold. Rotate 90°.

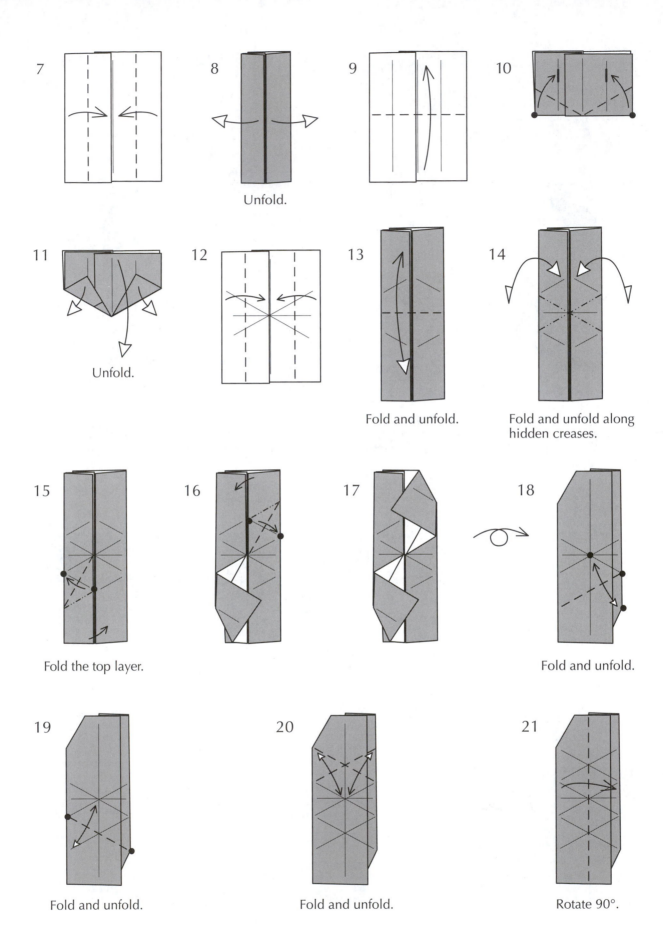

7

8

Unfold.

9

10

11

Unfold.

12

13

Fold and unfold.

14

Fold and unfold along hidden creases.

15

Fold the top layer.

16

17

18

Fold and unfold.

19

Fold and unfold.

20

Fold and unfold.

21

Rotate 90°.

22

Fold into the middle layer.

23

Reverse-fold.

24

Wrap around.

25

26

Repeat steps 23–24.

27

Fold all the layers together and unfold.

28

Fold all the layers together and unfold.

29

Open to view the inside.

30

Note the pocket. If the pocket is at the bottom, turn over and open again. Fold, then flatten.

31

tab

Puff out at the upper dot and at the similar dot behind. Tuck the tab into the pocket shown in the previous step. The lower dots will meet.

32

Tuck inside.

33

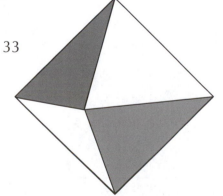

Duo-Colored Octahedron

Stellated Cube

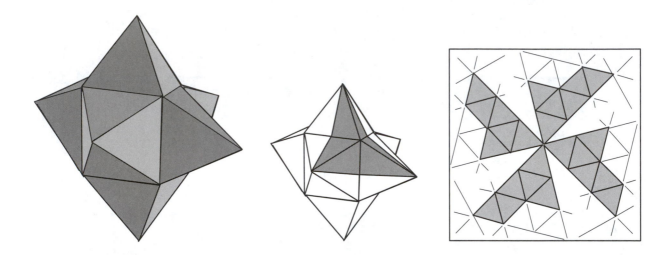

This model resembles eight pyramids, each on one face of a central cube. The layout shows square symmetry. The surface of this shape is the same as that of an octahedron. A side of the octahedron is represented by the shaded part in the middle drawing.

1

Fold and unfold.

2

Fold and unfold at the top and bottom.

3

Crease at the top and bottom.

4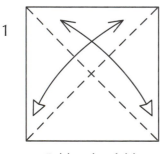

Unfold.

5

Fold and unfold at the top and bottom.

6

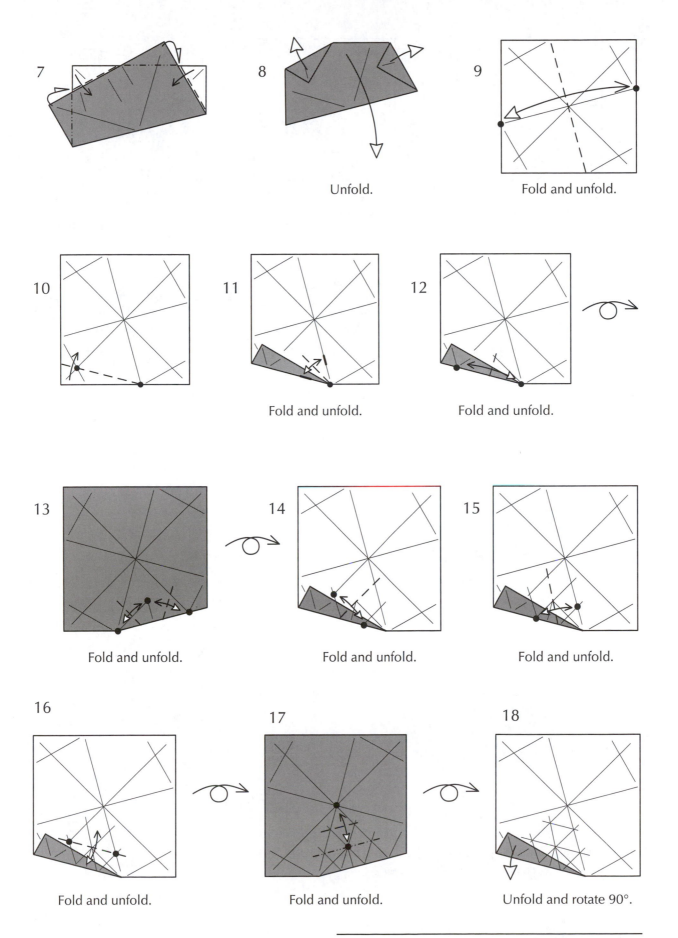

7

8

Unfold.

9

Fold and unfold.

10

11

Fold and unfold.

12

Fold and unfold.

13

Fold and unfold.

14

Fold and unfold.

15

Fold and unfold.

16

Fold and unfold.

17

Fold and unfold.

18

Unfold and rotate 90°.

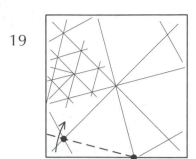

19

Repeat steps 10–18 three more times.

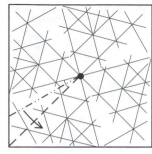

20

Bisect the angle and push in at the dot.

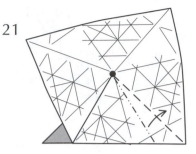

21

Repeat step 20 three more times. Rotate to view the outside so that the dot is at the bottom.

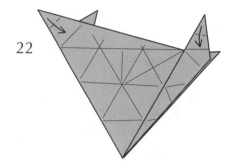

22

Repeat on all the corners.

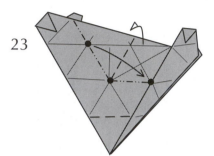

23

Puff out at the dot in the center.

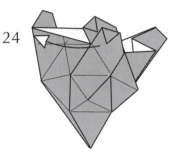

24

Unfold and rotate.

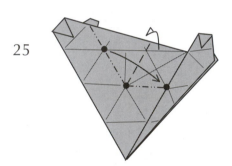

25

Repeat steps 23–24 three more times.

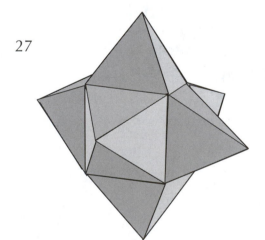

27

Stellated Cube

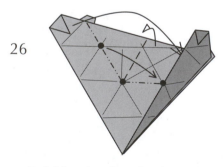

26

Refold and tuck under the paper shaded dark gray. Repeat three times.

Octahemioctahedron

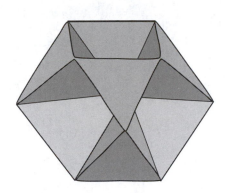

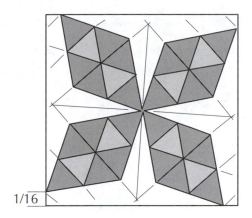

1/16

The octahemioctahedron comes from an octahedron where the six vertices are sunken. The structure is similar to that of the Octahedron, but the tab is 1/16 instead of 1/10. This makes for a larger model. The darker regions represent the sunken sides.

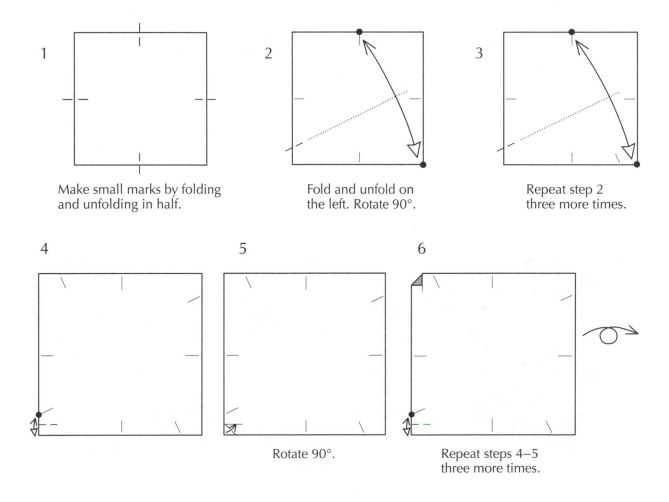

1. Make small marks by folding and unfolding in half.

2. Fold and unfold on the left. Rotate 90°.

3. Repeat step 2 three more times.

4.

5. Rotate 90°.

6. Repeat steps 4–5 three more times.

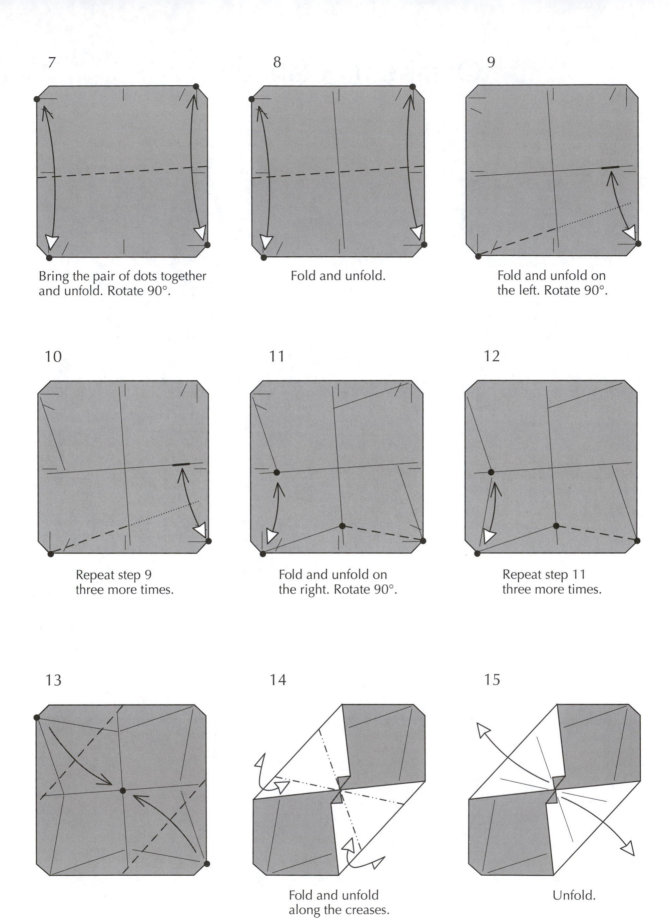

7

Bring the pair of dots together and unfold. Rotate 90°.

8

Fold and unfold.

9

Fold and unfold on the left. Rotate 90°.

10

Repeat step 9 three more times.

11

Fold and unfold on the right. Rotate 90°.

12

Repeat step 11 three more times.

13

14

Fold and unfold along the creases.

15

Unfold.

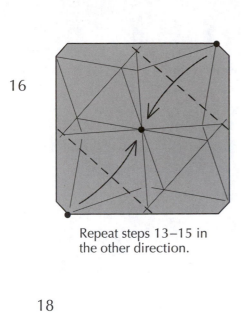

16

Repeat steps 13–15 in
the other direction.

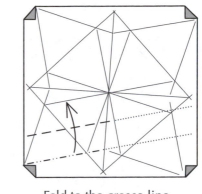

17

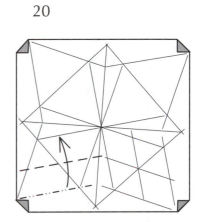

Fold to the crease line.
Only crease on the left.

18

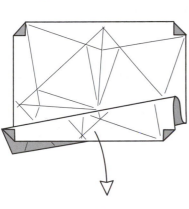

Unfold.

19

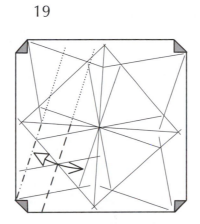

Fold to the crease line and
unfold. Only crease on the
bottom. Rotate 90°.

20

Repeat steps 17–19 three
more times. Rotate.

21

22

23

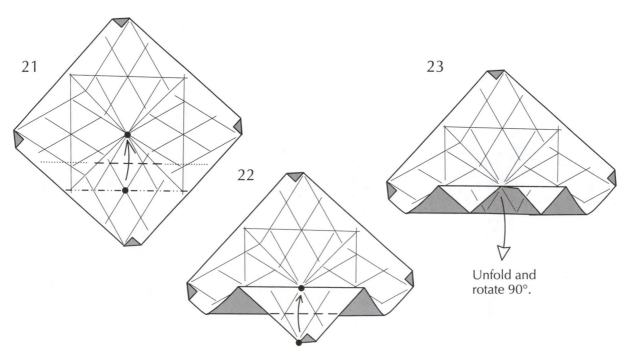

Unfold and
rotate 90°.

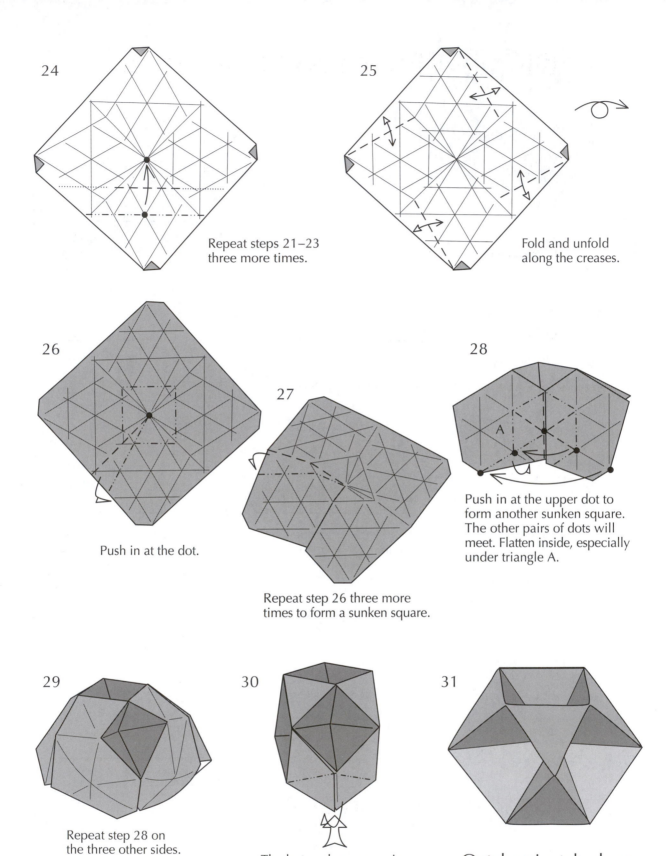

24

Repeat steps 21–23
three more times.

25

Fold and unfold
along the creases.

26

Push in at the dot.

27

Repeat step 26 three more
times to form a sunken square.

28

Push in at the upper dot to
form another sunken square.
The other pairs of dots will
meet. Flatten inside, especially
under triangle A.

29

Repeat step 28 on
the three other sides.

30

The last sunken square is
formed by four connected
reverse folds.

31

Octahemioctahedron

Dimpled Truncated Octahedron

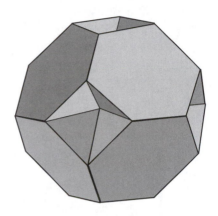

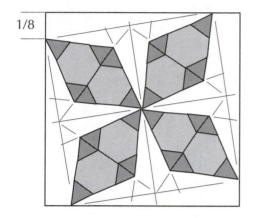

This design comes from the octahedron where the six vertices are sunken—but not all the way as in the octahemioctahedron. A tab of 1/8 is used. The darker regions in the crease pattern represent the sunken sides.

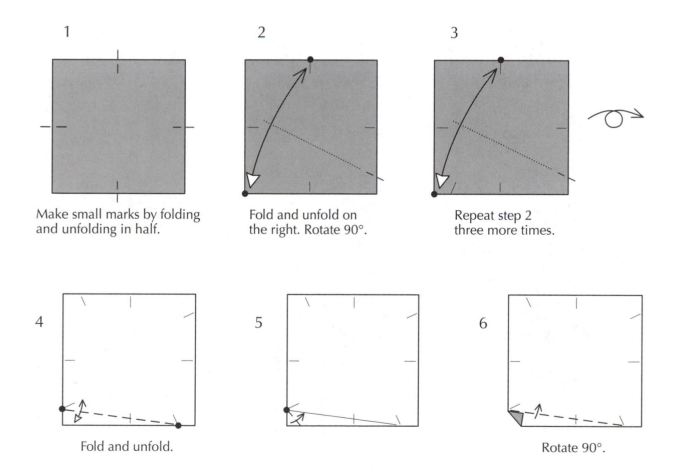

1

Make small marks by folding and unfolding in half.

2

Fold and unfold on the right. Rotate 90°.

3

Repeat step 2 three more times.

4

Fold and unfold.

5

6

Rotate 90°.

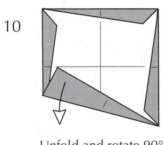

7

Repeat steps 4–6
three more times.

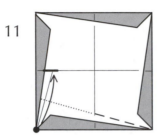

8

Fold and unfold.

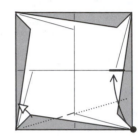

9

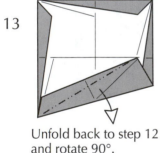

10

Unfold and rotate 90°.

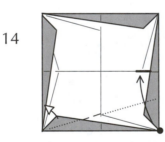

11

Repeat steps 9–10
three more times.

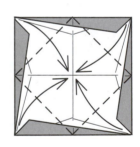

12

Slide the paper on the left so
that you only crease one layer.

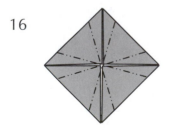

13

Unfold back to step 12
and rotate 90°.

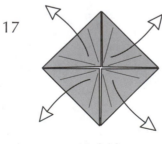

14

Repeat steps 12–13
three more times.

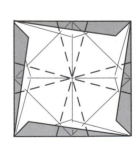

15

16

Fold and unfold all the
layers along the creases.

17

Unfold.

18

Fold and unfold along
the creases. Rotate.

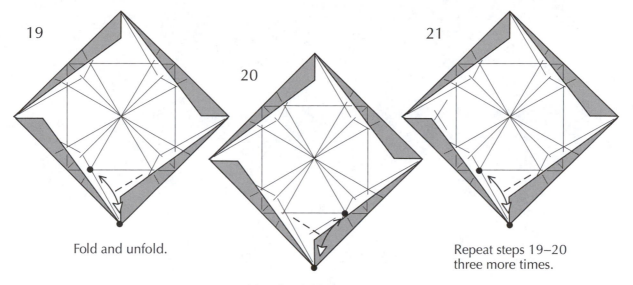

19

Fold and unfold.

20

Fold and unfold. Rotate 90°.

21

Repeat steps 19–20 three more times.

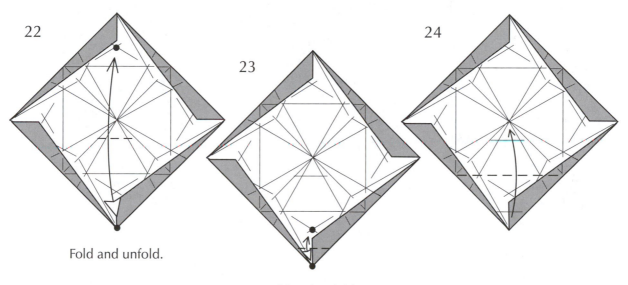

22

Fold and unfold.

23

Fold and unfold.

24

25

Fold and unfold.

26

27

Unfold and rotate 90°.

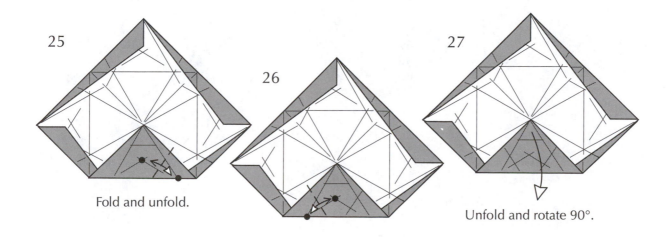

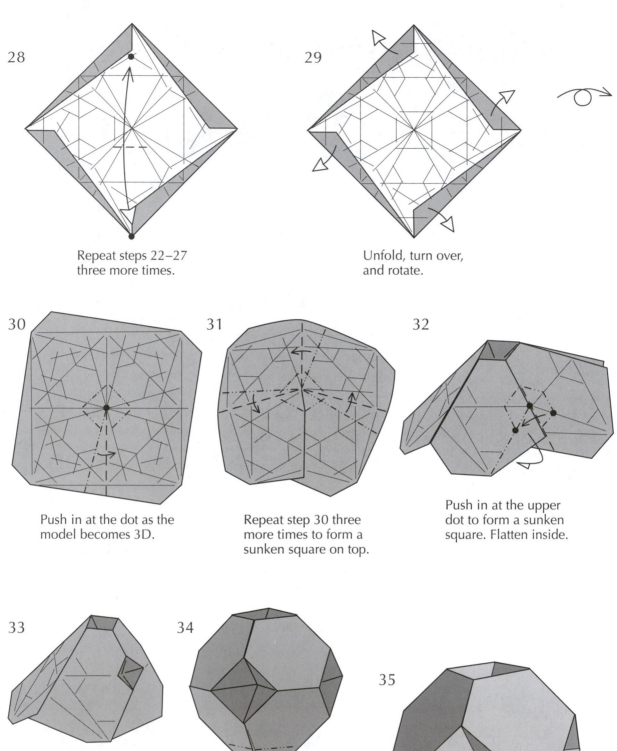

28

Repeat steps 22–27
three more times.

29

Unfold, turn over,
and rotate.

30

Push in at the dot as the
model becomes 3D.

31

Repeat step 30 three
more times to form a
sunken square on top.

32

Push in at the upper
dot to form a sunken
square. Flatten inside.

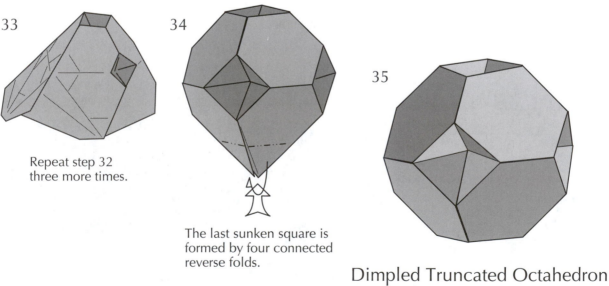

33

Repeat step 32
three more times.

34

The last sunken square is
formed by four connected
reverse folds.

35

Dimpled Truncated Octahedron

More Platonic Solids Design

The five Platonic solids are convex, have faces that are identical regular polygons, and have identical vertices. We have seen three of them so far, the tetrahedron, cube, and octahedron. The remaining two, the icosahedron and dodecahedron, are shown here.

Icosahedron

The icosahedron is composed of 20 equilateral triangles. Here are a few possible layouts and crease patterns. Several layouts have a band of ten triangles.

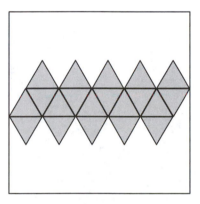

Band of 10 triangles with 5 above and 5 below.

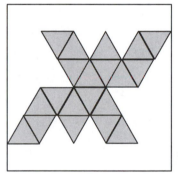

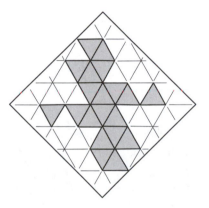

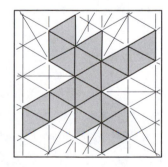

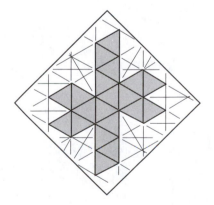

This version is diagrammed.

Dodecahedron

Designing an origami dodecahedron is quite a challenge. The dodecahedron is composed of 12 pentagons. Here are a few layouts.

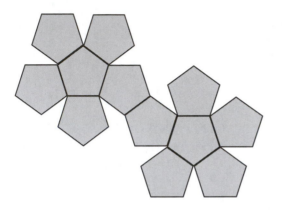

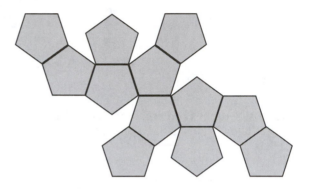

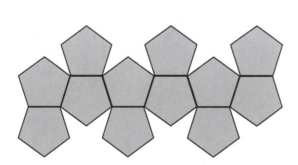

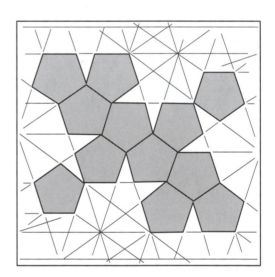

This crease pattern is diagrammed.

The Five Platonic Solids

| Tetrahedron | Cube | Octahedron | Icosahedron | Dodecahedron |

Icosahedron

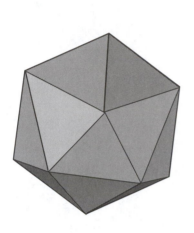

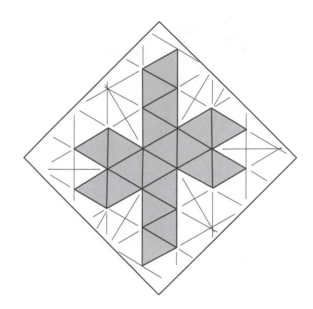

The icosahedron is composed of 20 equilateral triangles. The layout shows a band of ten triangles down the diagonal with five triangles on each side. Odd symmetry is used.

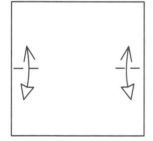

1

Fold and unfold on the left and right.

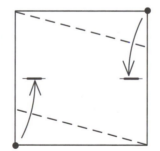

2

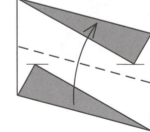

3

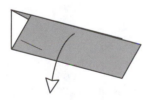

4

Unfold.

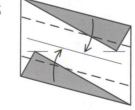

5

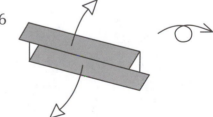

6

Unfold.

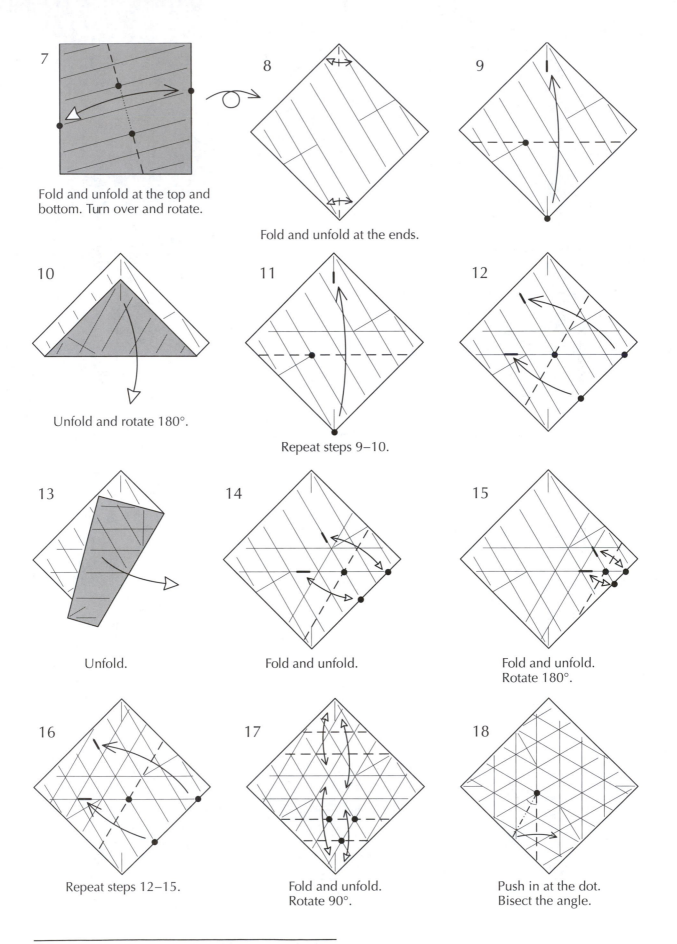

7 Fold and unfold at the top and bottom. Turn over and rotate.

8 Fold and unfold at the ends.

9

10 Unfold and rotate 180°.

11 Repeat steps 9–10.

12

13 Unfold.

14 Fold and unfold.

15 Fold and unfold. Rotate 180°.

16 Repeat steps 12–15.

17 Fold and unfold. Rotate 90°.

18 Push in at the dot. Bisect the angle.

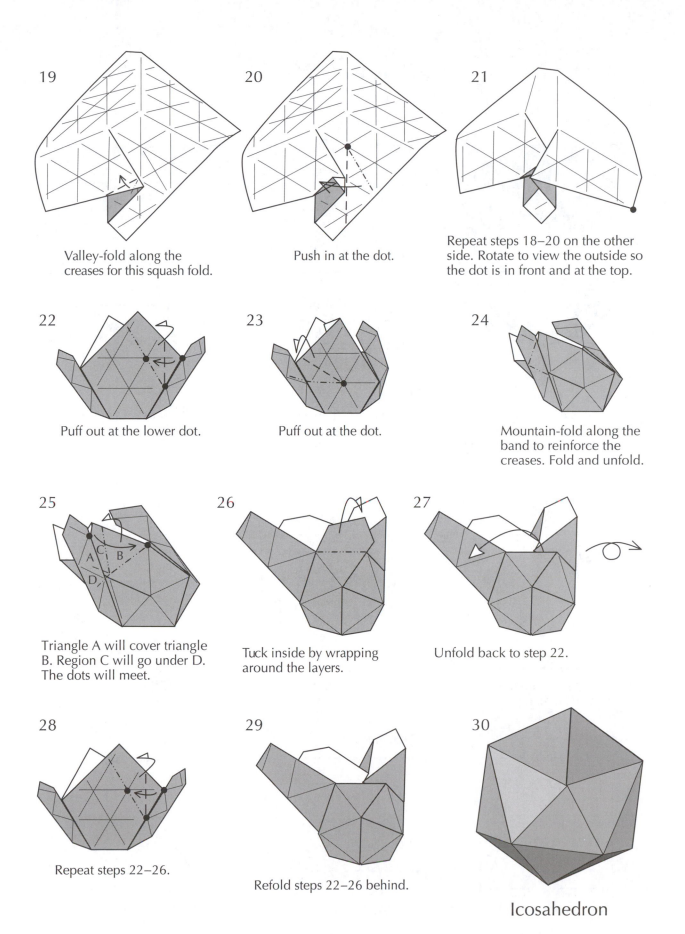

19

Valley-fold along the creases for this squash fold.

20

Push in at the dot.

21

Repeat steps 18–20 on the other side. Rotate to view the outside so the dot is in front and at the top.

22

Puff out at the lower dot.

23

Puff out at the dot.

24

Mountain-fold along the band to reinforce the creases. Fold and unfold.

25

Triangle A will cover triangle B. Region C will go under D. The dots will meet.

26

Tuck inside by wrapping around the layers.

27

Unfold back to step 22.

28

Repeat steps 22–26.

29

Refold steps 22–26 behind.

30

Icosahedron

Dodecahedron

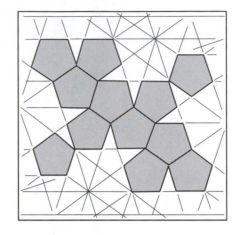

To Plato, this dodecahedron, the quintessence (the "fifth being"), represented the whole universe. The dodecahedron has 12 pentagonal faces. The crease pattern shows odd symmetry with a line going through the center at 36° from the horizontal line.

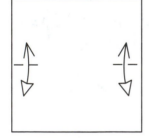

1. Fold and unfold on the left and right.

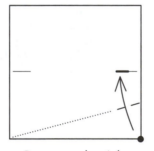

2. Crease on the right.

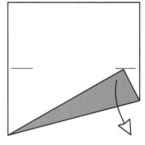

3. Unfold and rotate 180°.

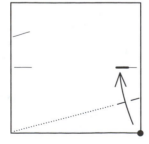

4. Repeat steps 2–3.

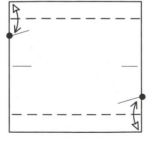

5. Fold and unfold.

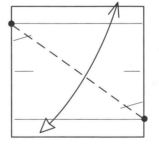

6. Fold and unfold.

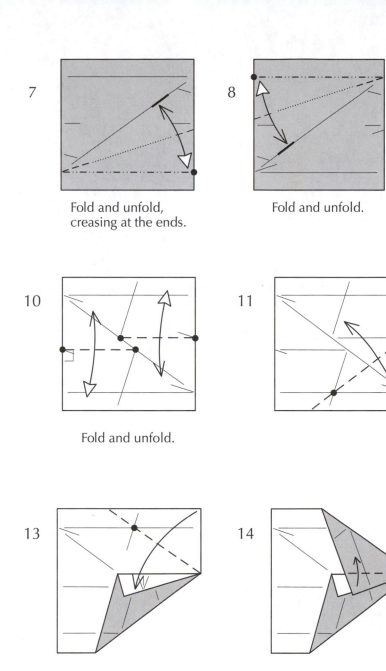

7

Fold and unfold,
creasing at the ends.

8

Fold and unfold.

9

Fold and unfold.

10

Fold and unfold.

11

12

Fold along a
hidden crease.

13

14

Fold along a hidden crease.

15

Unfold and rotate 180°.

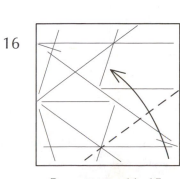

16

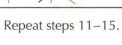

Repeat steps 11–15.

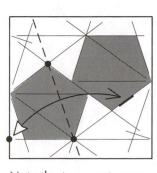

17

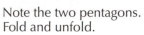

Note the two pentagons.
Fold and unfold.

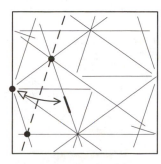

18

Fold and unfold.

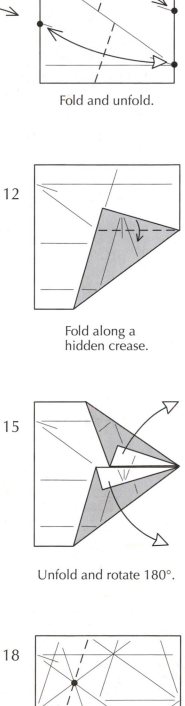

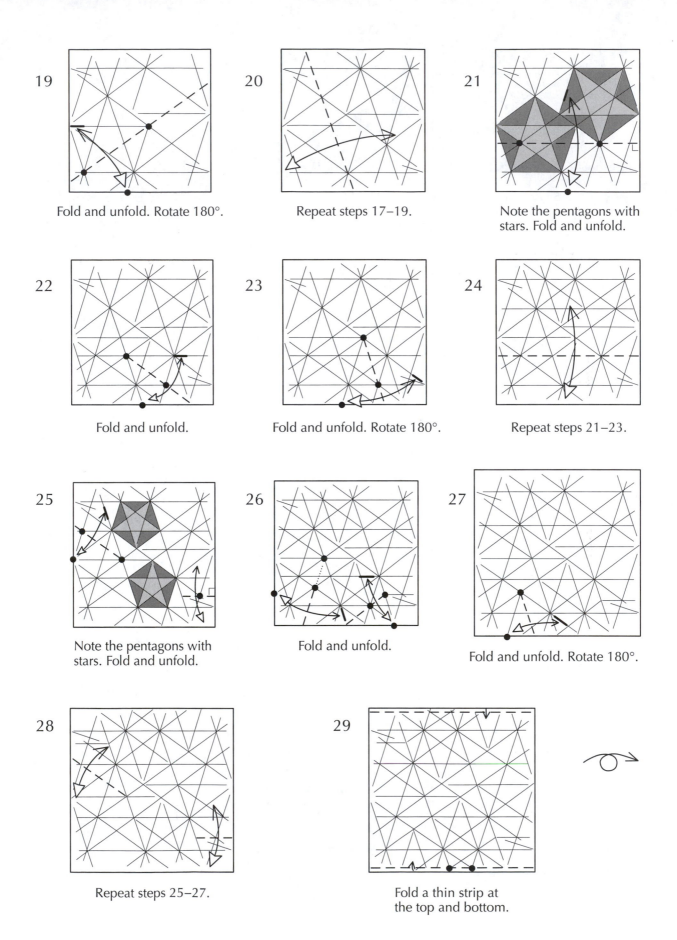

19 Fold and unfold. Rotate 180°.

20 Repeat steps 17–19.

21 Note the pentagons with stars. Fold and unfold.

22 Fold and unfold.

23 Fold and unfold. Rotate 180°.

24 Repeat steps 21–23.

25 Note the pentagons with stars. Fold and unfold.

26 Fold and unfold.

27 Fold and unfold. Rotate 180°.

28 Repeat steps 25–27.

29 Fold a thin strip at the top and bottom.

30

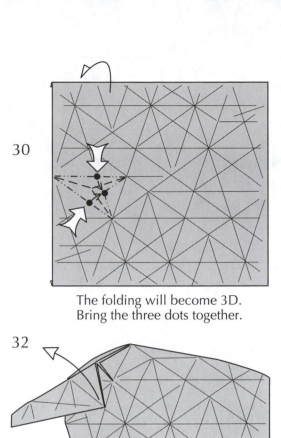

The folding will become 3D.
Bring the three dots together.

31

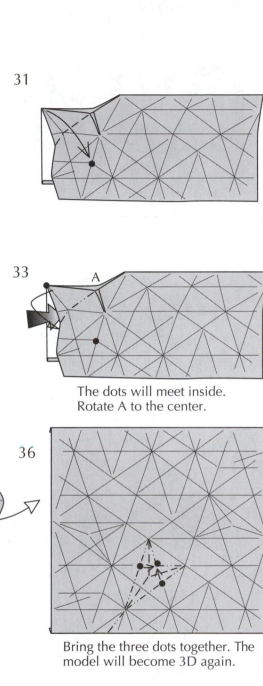

32

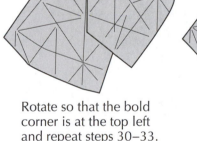

Unfold.

33

The dots will meet inside.
Rotate A to the center.

34

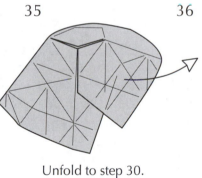

Rotate so that the bold
corner is at the top left
and repeat steps 30–33.

35

Unfold to step 30.

36

Bring the three dots together. The
model will become 3D again.

37

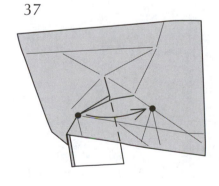

38

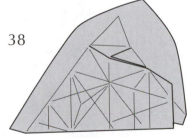

Rotate and repeat steps
36–37 on the opposite edge.

39

Unfold to step 36.

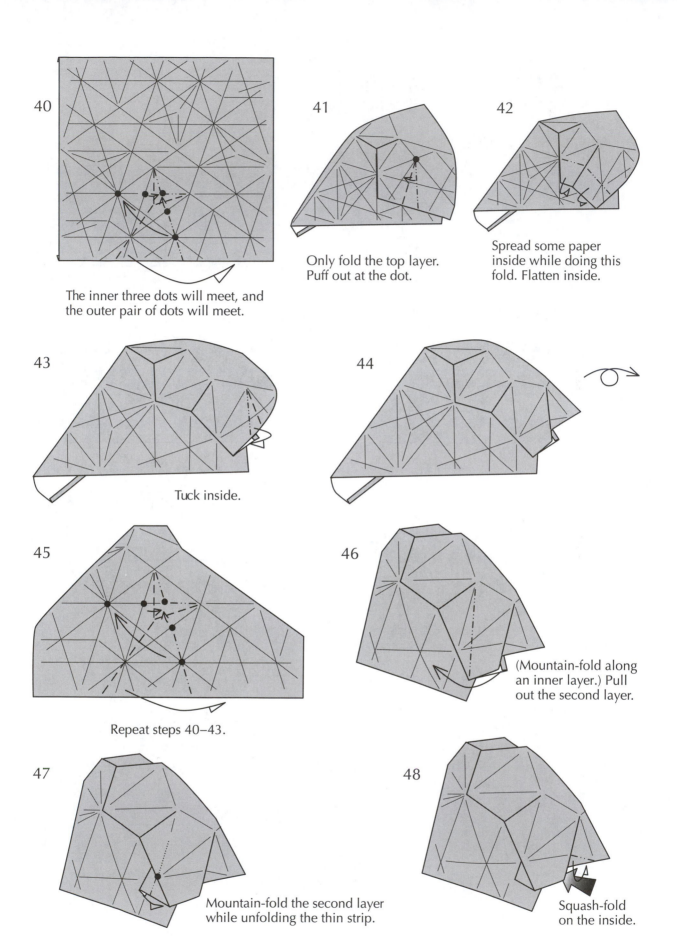

40

The inner three dots will meet, and the outer pair of dots will meet.

41

Only fold the top layer. Puff out at the dot.

42

Spread some paper inside while doing this fold. Flatten inside.

43

Tuck inside.

44

45

Repeat steps 40–43.

46

(Mountain-fold along an inner layer.) Pull out the second layer.

47

Mountain-fold the second layer while unfolding the thin strip.

48

Squash-fold on the inside.

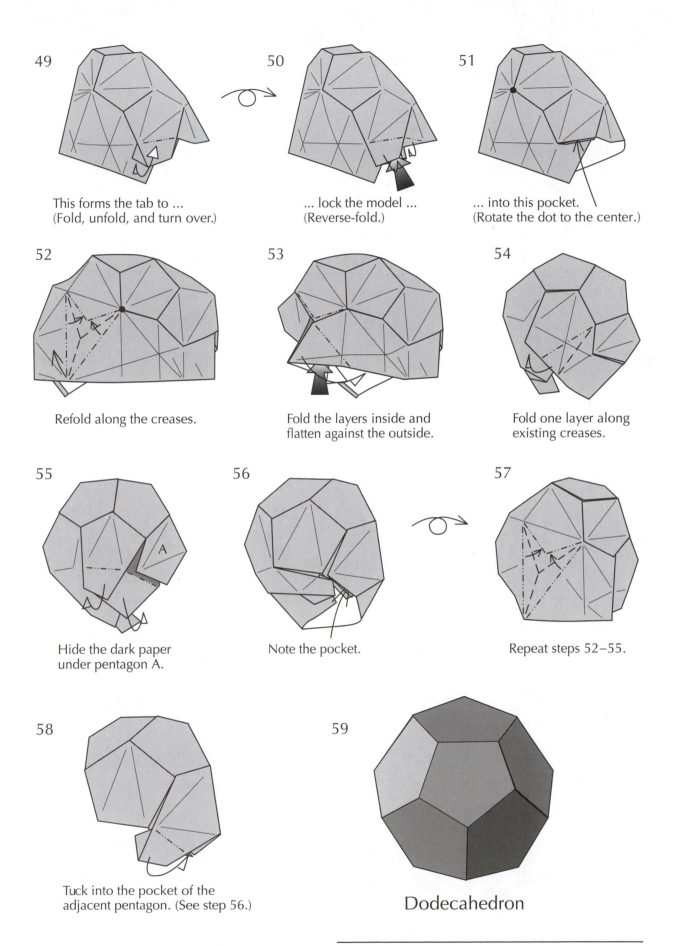

49 This forms the tab to ...
(Fold, unfold, and turn over.)

50 ... lock the model ...
(Reverse-fold.)

51 ... into this pocket.
(Rotate the dot to the center.)

52 Refold along the creases.

53 Fold the layers inside and
flatten against the outside.

54 Fold one layer along
existing creases.

55 Hide the dark paper
under pentagon A.

56 Note the pocket.

57 Repeat steps 52–55.

58 Tuck into the pocket of the
adjacent pentagon. (See step 56.)

59 Dodecahedron

Sunken Platonic Solids Design

A sunken version of a polyhedron is created by taking the center of each face and sinking it toward the center of the polyhedron. This replaces each *n*-sided face with *n* triangular faces. Sunken polyhedra are concave. In general, sunken models hold and lock better than the convex ones.

The group of sunken Platonic solids is interesting. It takes some unusual folds to capture their shapes.

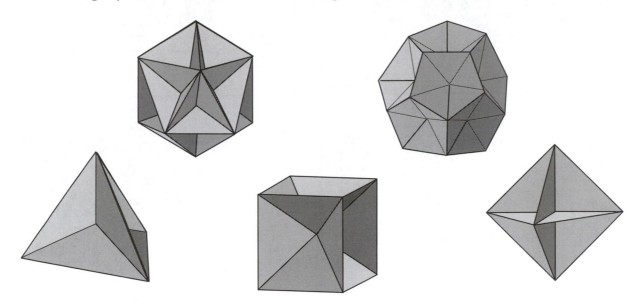

Sunken Octahedron

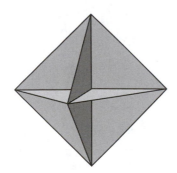

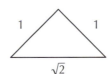

Each side is an isosceles triangle with an apex angle of 90°.

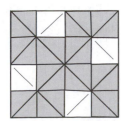

A convenient layout has square symmetry. This crease pattern produces the largest model but has no tab.

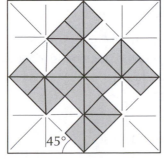

Rotating by 45° produces a convenient crease pattern, but the model becomes small and thick.

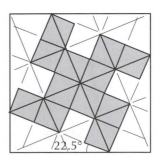

By rotating 22.5°, there is enough tab while not compromising the size of the final model as much. This version is diagrammed.

Sunken Tetrahedron

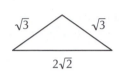

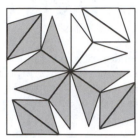

Using 3/4 square symmetry and filling the square well, this version is the most efficient for use of paper. This version is diagrammed.

The sunken tetrahedron is composed of 24 isosceles triangles. The dimensions of each side is shown above.

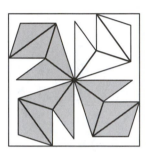

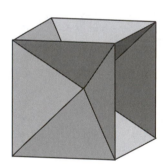

More samples of 3/4 square symmetry.

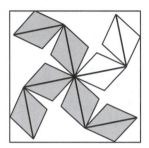

Even symmetry.

Sunken Cube

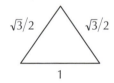

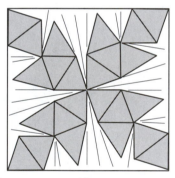

Crease pattern with square symmetry.

The sunken cube is composed of 24 isosceles triangles. The length of one side of those triangles is equal to half the length of the diagonal through the center of the cube. For a 1 × 1 × 1 cube, the length of the diagonal on a face (dotted line) is $\sqrt{2}$ (by the Pythagorean Theorem). So, the length of the diagonal through the center of the cube is $\sqrt{3}$, and the length of the side on the triangular face is $\sqrt{3}/2$.

The dots in the sunken cube define the triangle to the right.

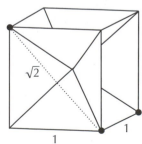

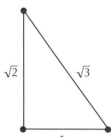

Sunken Dodecahedron

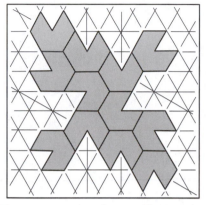

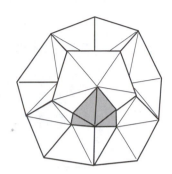

The crease pattern shows odd symmetry.

This sunken dodecahedron is composed of 60 equilateral triangles. Difficult as it is, using equilateral triangles does simplify the design and folding. This model has the same surface as the icosahedron. In the third drawing, the gray sides represent one side of the icosahedron. There are two other related shapes with the same surface: one is a dimpled truncated icosahedron, the other is a dimpled soccer ball.

Sunken Icosahedron

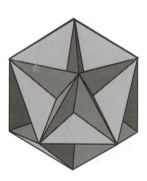

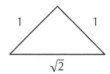

Each side is an isosceles triangle with an apex angle of 90°.

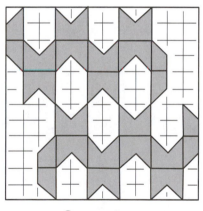

Crease pattern.

This shape is so complex I could not find a useful two-dimensional layout. But I imagined how a fan (step 22) can turn into this model. The layout begins by dividing into twelfths. Also, because of its complexity, the folded model has a rugged beauty.

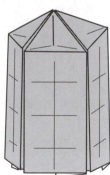

A fan turns into a tower to create the sunken icosahedron.

Sunken Octahedron

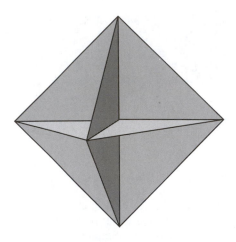

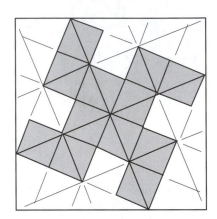

All the faces meet at the center, and this model can be viewed as three intersecting squares corresponding to the x, y, and z planes. A polyhedron with zero volume, like this one, can be called a nolid ("not a solid"). This would be an octahedral nolid.

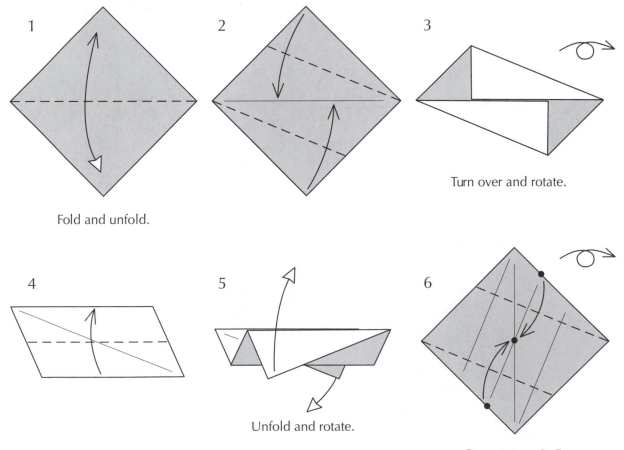

1

Fold and unfold.

2

3

Turn over and rotate.

4

5

Unfold and rotate.

6

Repeat steps 2–5.

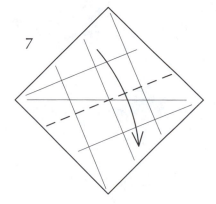

7

Fold along the
crease and rotate.

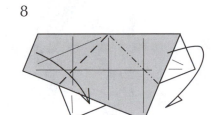

8

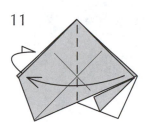

9

Open and follow the dot.

10

Turn over and repeat.

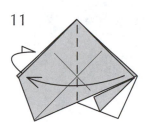

11

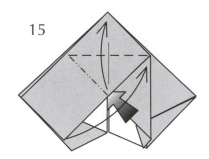

12

Turn over and repeat.

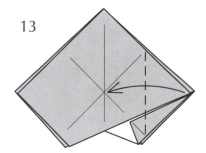

13

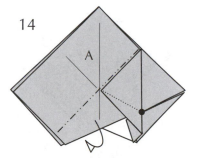

14

Mountain-fold along the
crease, sliding the hidden
corner at the dot under area A.

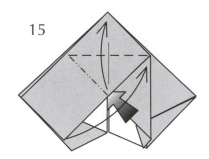

15

Squash-fold.

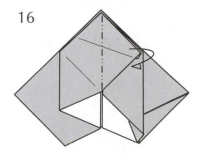

16

Fold behind.

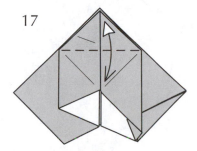

17

Fold and unfold.

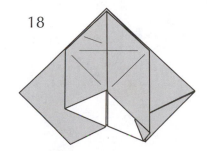

18

Turn over and
repeat steps 13–17.

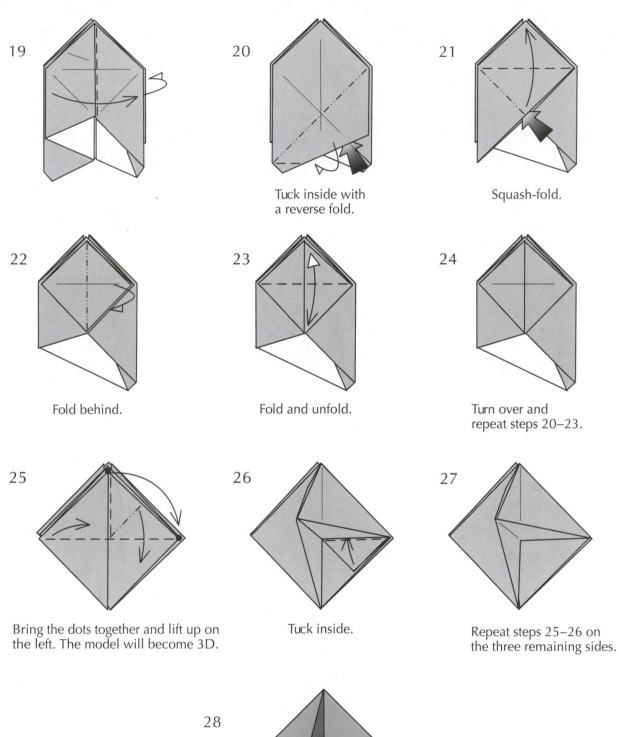

19

20

Tuck inside with
a reverse fold.

21

Squash-fold.

22

Fold behind.

23

Fold and unfold.

24

Turn over and
repeat steps 20–23.

25

Bring the dots together and lift up on
the left. The model will become 3D.

26

Tuck inside.

27

Repeat steps 25–26 on
the three remaining sides.

28

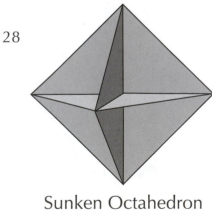

Sunken Octahedron

Sunken Tetrahedron

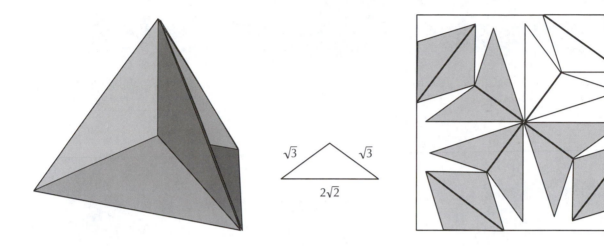

√3   √3

2√2

The sunken tetrahedron is composed of twelve isosceles triangles all meeting in the center. The sides of each triangle are proportional to $2\sqrt{2}$, $\sqrt{3}$, $\sqrt{3}$. This can also be called a tetrahedral nolid. The crease pattern shows 3/4 square symmetry.

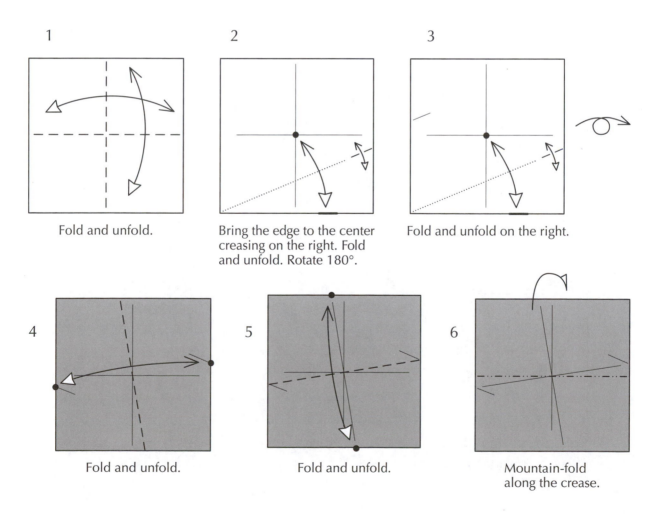

1

Fold and unfold.

2

Bring the edge to the center creasing on the right. Fold and unfold. Rotate 180°.

3

Fold and unfold on the right.

4

Fold and unfold.

5

Fold and unfold.

6

Mountain-fold along the crease.

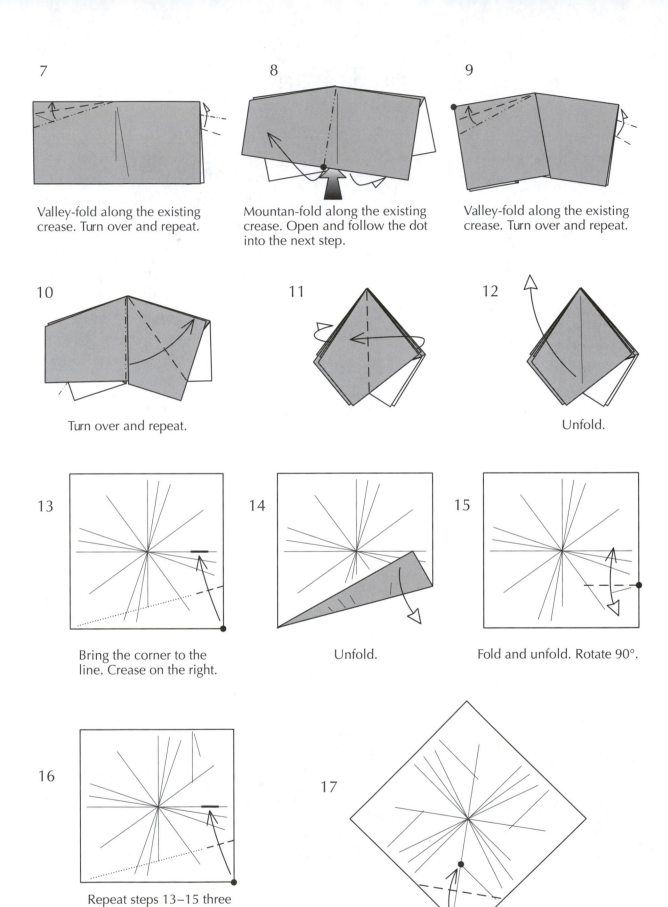

7

Valley-fold along the existing crease. Turn over and repeat.

8

Mountan-fold along the existing crease. Open and follow the dot into the next step.

9

Valley-fold along the existing crease. Turn over and repeat.

10

Turn over and repeat.

11

12

Unfold.

13

Bring the corner to the line. Crease on the right.

14

Unfold.

15

Fold and unfold. Rotate 90°.

16

Repeat steps 13–15 three more times. Rotate.

17

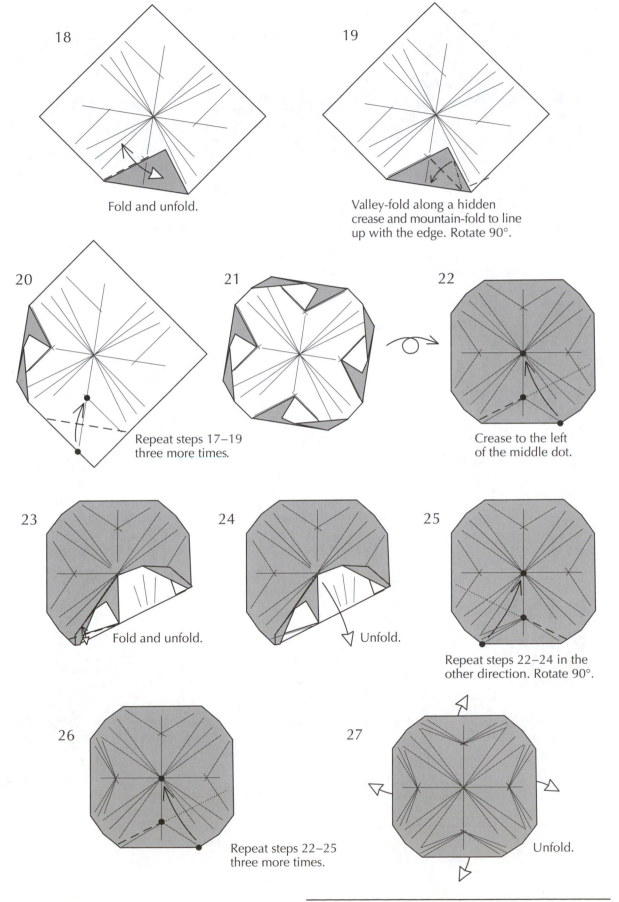

18

Fold and unfold.

19

Valley-fold along a hidden crease and mountain-fold to line up with the edge. Rotate 90°.

20

Repeat steps 17–19 three more times.

21

22

Crease to the left of the middle dot.

23

Fold and unfold.

24

Unfold.

25

Repeat steps 22–24 in the other direction. Rotate 90°.

26

Repeat steps 22–25 three more times.

27

Unfold.

28

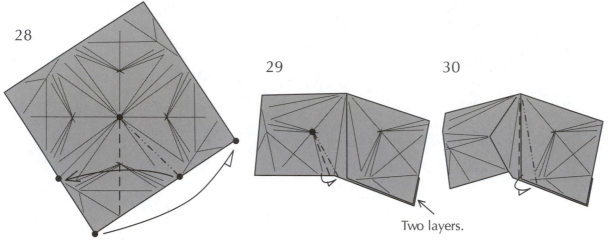

Puff out at the upper dot as
the model becomes 3D.

Push in at the dot.
Note the two layers.

Two layers.

31

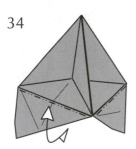

Push in at the dot.

32

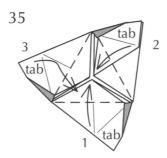

Fold all the layers
behind on the crease.

33

Repeat steps 29–32 two more
times beginning on the right
and cycling around.

34

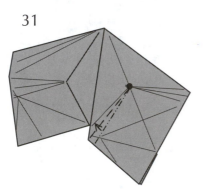

Fold and unfold on three
sides. Rotate to view the
bottom.

35

Fold paper 1, then fold 2
which covers tab 1. Then
fold 3 to cover tab 2 and
tuck tab 3 under 1. Push
in at the center. Rotate
this side to the bottom.

36

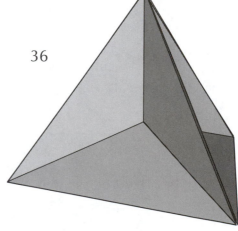

Sunken Tetrahedron

Sunken Cube

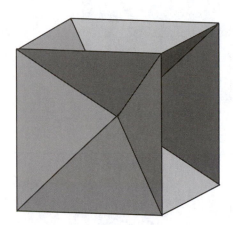

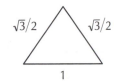

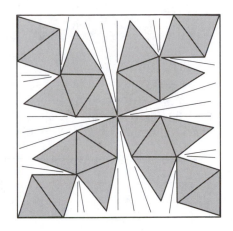

The sunken cube is composed of 24 isosceles triangles all meeting at the center. The sides of each triangle are proportional to 1, $\sqrt{3}/2$, $\sqrt{3}/2$. This beautiful shape can also be called a hexahedral nolid, that is, it comes from the cube but has no volume. The crease pattern shows square symmetry.

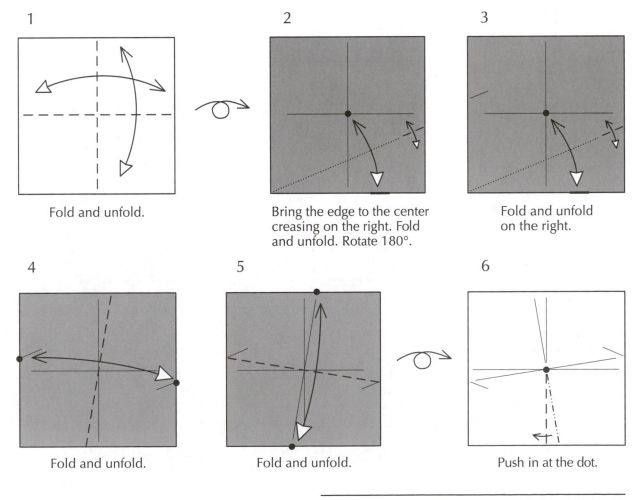

1

Fold and unfold.

2

Bring the edge to the center creasing on the right. Fold and unfold. Rotate 180°.

3

Fold and unfold on the right.

4

Fold and unfold.

5

Fold and unfold.

6

Push in at the dot.

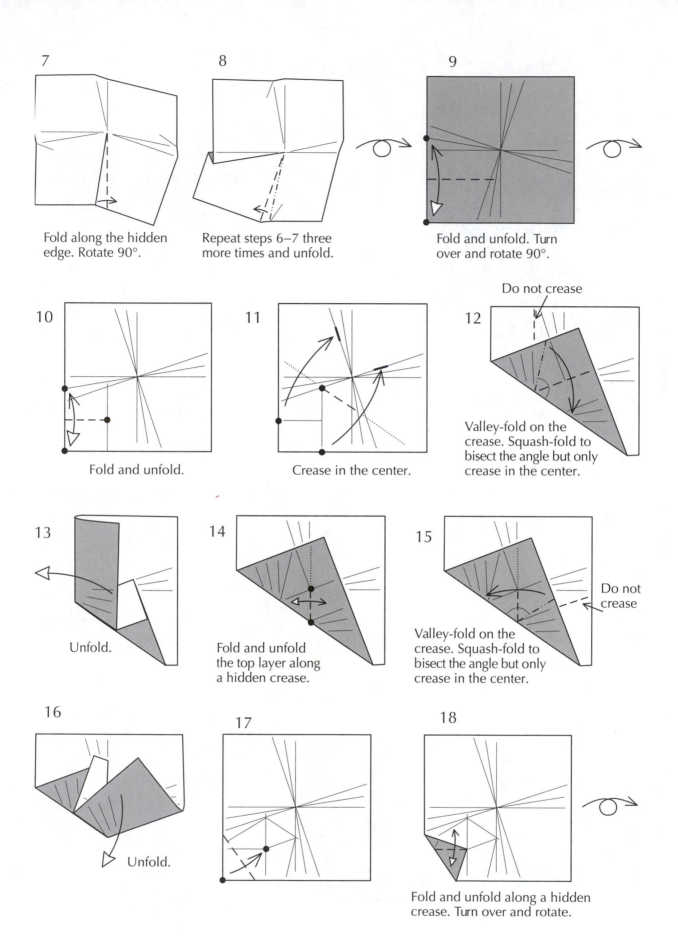

7

Fold along the hidden edge. Rotate 90°.

8

Repeat steps 6–7 three more times and unfold.

9

Fold and unfold. Turn over and rotate 90°.

10

Fold and unfold.

11

Crease in the center.

12

Do not crease

Valley-fold on the crease. Squash-fold to bisect the angle but only crease in the center.

13

Unfold.

14

Fold and unfold the top layer along a hidden crease.

15

Do not crease

Valley-fold on the crease. Squash-fold to bisect the angle but only crease in the center.

16

Unfold.

17

18

Fold and unfold along a hidden crease. Turn over and rotate.

19

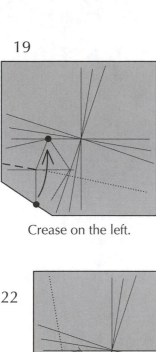

Crease on the left.

20

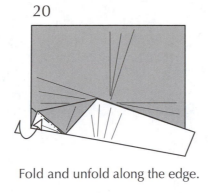

Fold and unfold along the edge.

21

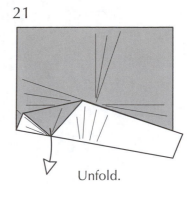

Unfold.

22

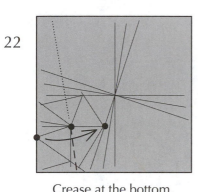

Crease at the bottom.

23

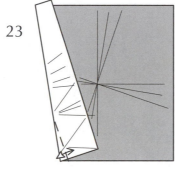

Fold and unfold
both layers.

24

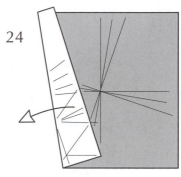

Unfold and rotate 90°.

25

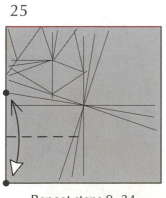

Repeat steps 9–24
three more times.

26

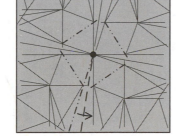

Push in at the dot as the
model becomes 3D.

27

Form a sunken square on top.
Rotate the top to the bottom.

28

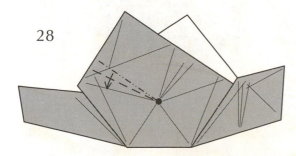

Push in at the dot.

29

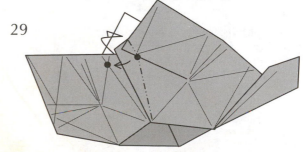

Slide the paper. There is a
small squash fold at the top.

30

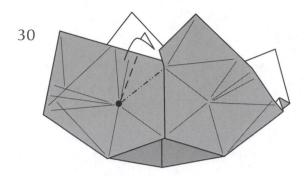

Push in at the dot.

31

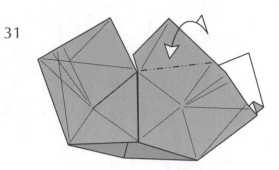

Fold and unfold.

32

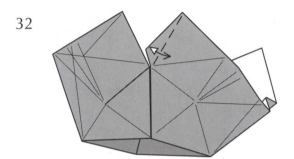

Fold and unfold.

33

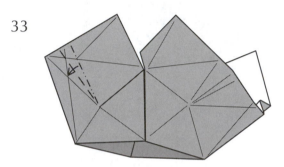

Repeat steps 28–32
three more times.

34

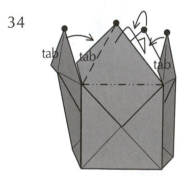

Fold the four corners inside to
form the last sunken square.
Interlock to cover the tabs.

35

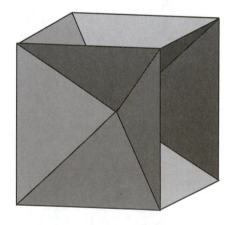

Sunken Cube

Sunken Dodecahedron

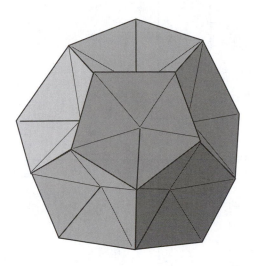

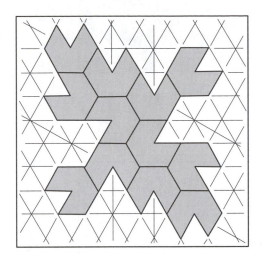

This sunken dodecahedron, one of several stellated icosahedrons, is composed of 60 equilateral triangles. This version has the same surface as an icosahedron. The crease pattern shows odd symmetry.

1

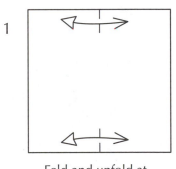

Fold and unfold at the top and bottom.

2

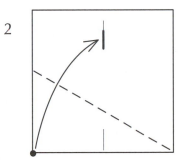

3

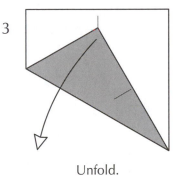

Unfold.

4

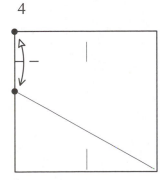

Fold and unfold on the left. Rotate 180°.

5

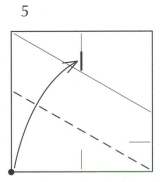

Repeat steps 2–4.

6

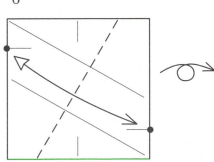

Fold and unfold.

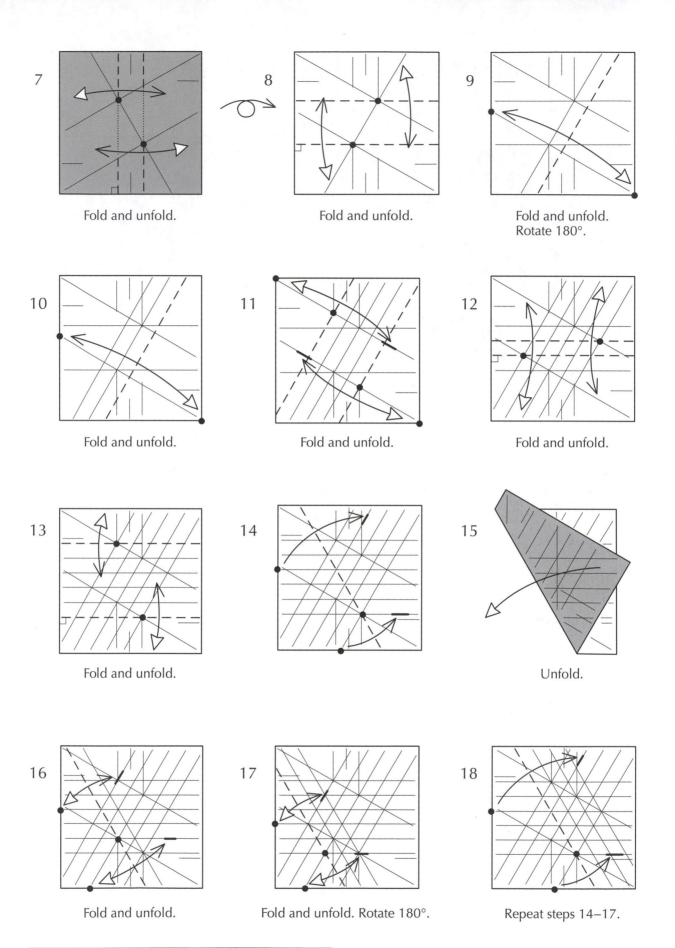

7 Fold and unfold.

8 Fold and unfold.

9 Fold and unfold. Rotate 180°.

10 Fold and unfold.

11 Fold and unfold.

12 Fold and unfold.

13 Fold and unfold.

14

15 Unfold.

16 Fold and unfold.

17 Fold and unfold. Rotate 180°.

18 Repeat steps 14–17.

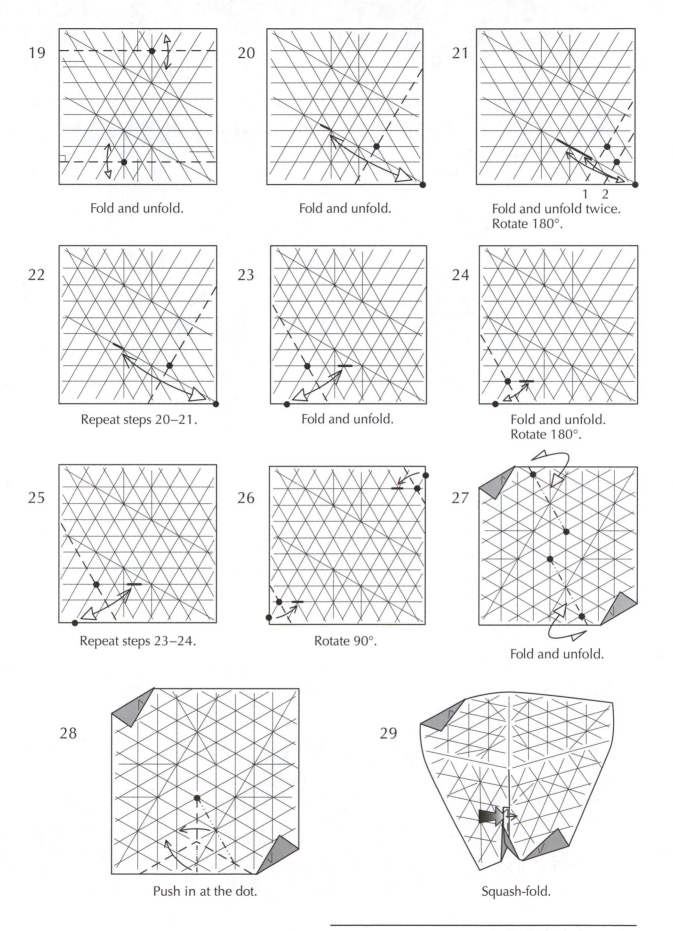

19   Fold and unfold.

20   Fold and unfold.

21   Fold and unfold twice.
Rotate 180°.

22   Repeat steps 20–21.

23   Fold and unfold.

24   Fold and unfold.
Rotate 180°.

25   Repeat steps 23–24.

26   Rotate 90°.

27   Fold and unfold.

28   Push in at the dot.

29   Squash-fold.

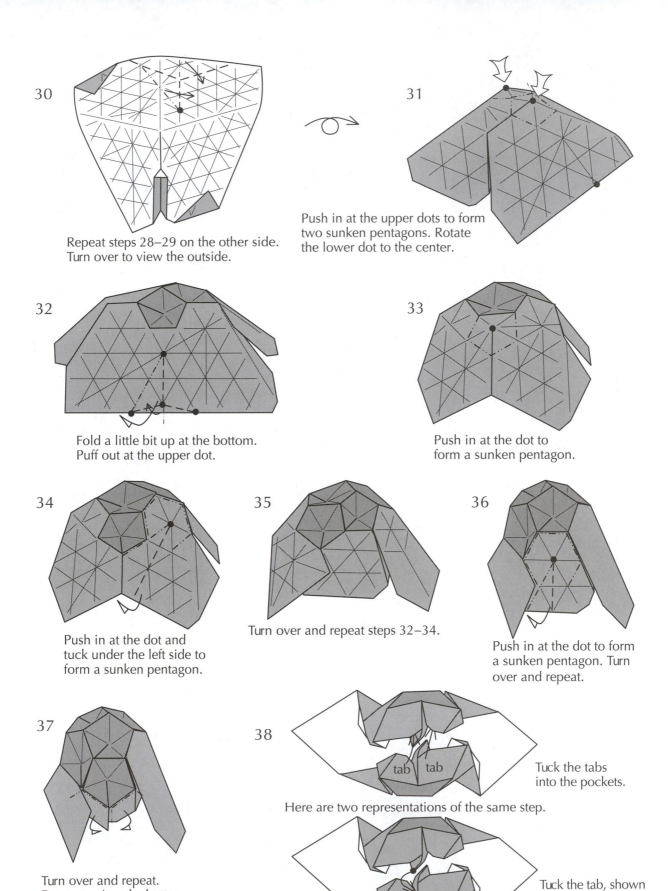

30

Repeat steps 28–29 on the other side.
Turn over to view the outside.

31

Push in at the upper dots to form
two sunken pentagons. Rotate
the lower dot to the center.

32

Fold a little bit up at the bottom.
Puff out at the upper dot.

33

Push in at the dot to
form a sunken pentagon.

34

Push in at the dot and
tuck under the left side to
form a sunken pentagon.

35

Turn over and repeat steps 32–34.

36

Push in at the dot to form
a sunken pentagon. Turn
over and repeat.

37

Turn over and repeat.
Rotate to view the bottom.

38

tab   tab

Here are two representations of the same step.

Tuck the tabs
into the pockets.

Tuck the tab, shown
with the dot, into
the pocket.

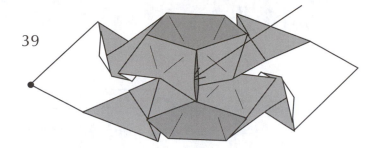

39

Note the pocket. This lock will loosen over the next few steps. Be sure to put it back by step 43. Rotate to view the outside so the dot is on the top at the front.

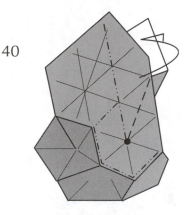

40

Push in at the dot to form a sunken pentagon. Turn over and repeat.

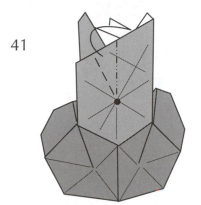

41

Push in at the dot and tuck inside. Turn over and repeat.

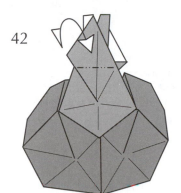

42

Fold and unfold. Turn over and repeat.

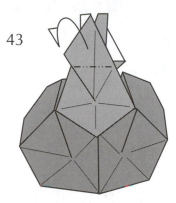

43

Tuck inside the pocket. Turn over and repeat.

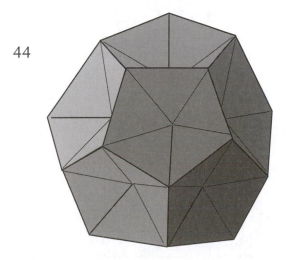

44

Sunken Dodecahedron

Sunken Icosahedron

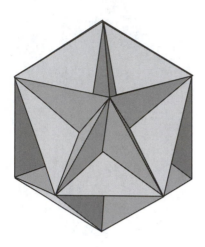

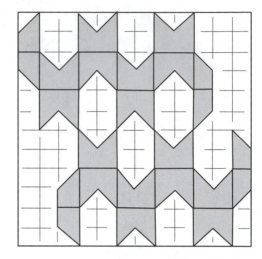

This complex shape is composed of 60 isosceles right triangles. It is formed by first making a fan. I gave it a four-star (very complex) rating because of the step sequence 20–33 as the fan is turned into a tower, but the folds before and after are not so difficult. The paper is divided into twelfths.

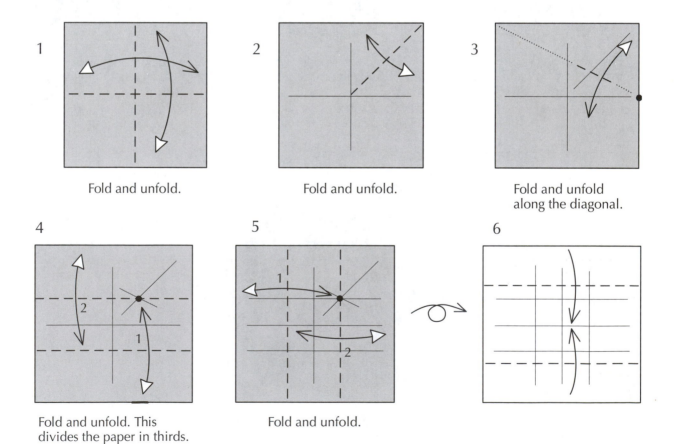

1 Fold and unfold.

2 Fold and unfold.

3 Fold and unfold along the diagonal.

4 Fold and unfold. This divides the paper in thirds.

5 Fold and unfold.

6

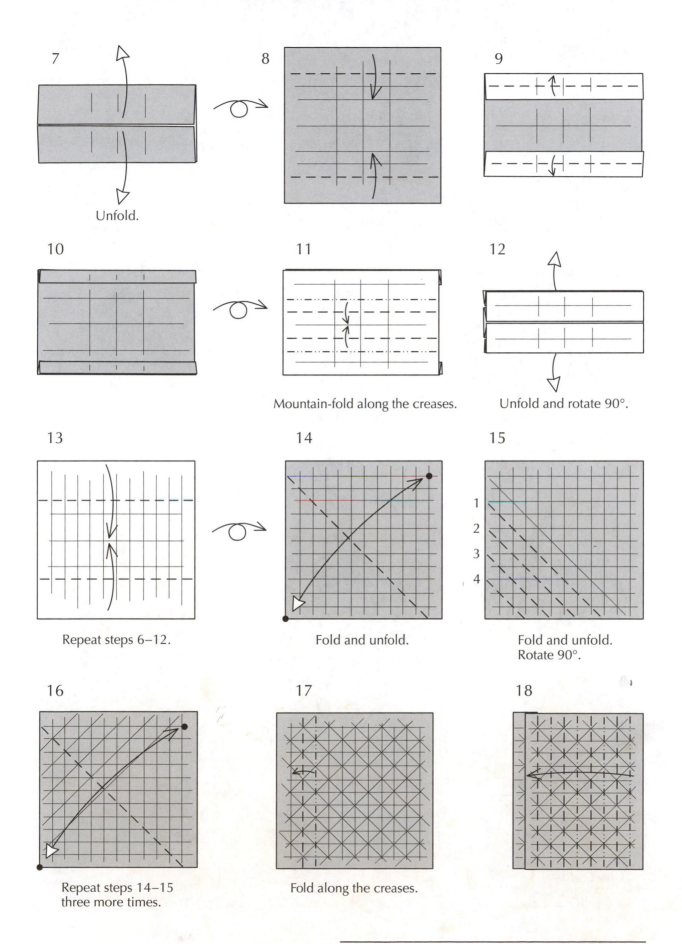

7

Unfold.

8

9

10

11

Mountain-fold along the creases.

12

Unfold and rotate 90°.

13

Repeat steps 6–12.

14

Fold and unfold.

15

1
2
3
4

Fold and unfold.
Rotate 90°.

16

Repeat steps 14–15
three more times.

17

Fold along the creases.

18

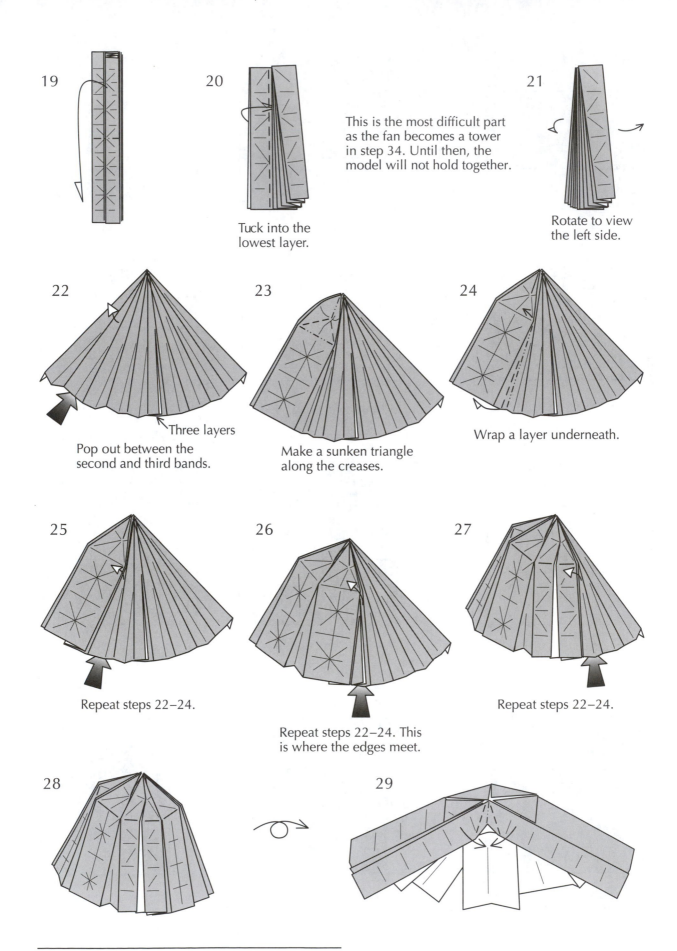

19

20

Tuck into the lowest layer.

This is the most difficult part as the fan becomes a tower in step 34. Until then, the model will not hold together.

21

Rotate to view the left side.

22

Three layers

Pop out between the second and third bands.

23

Make a sunken triangle along the creases.

24

Wrap a layer underneath.

25

Repeat steps 22–24.

26

Repeat steps 22–24. This is where the edges meet.

27

Repeat steps 22–24.

28

29

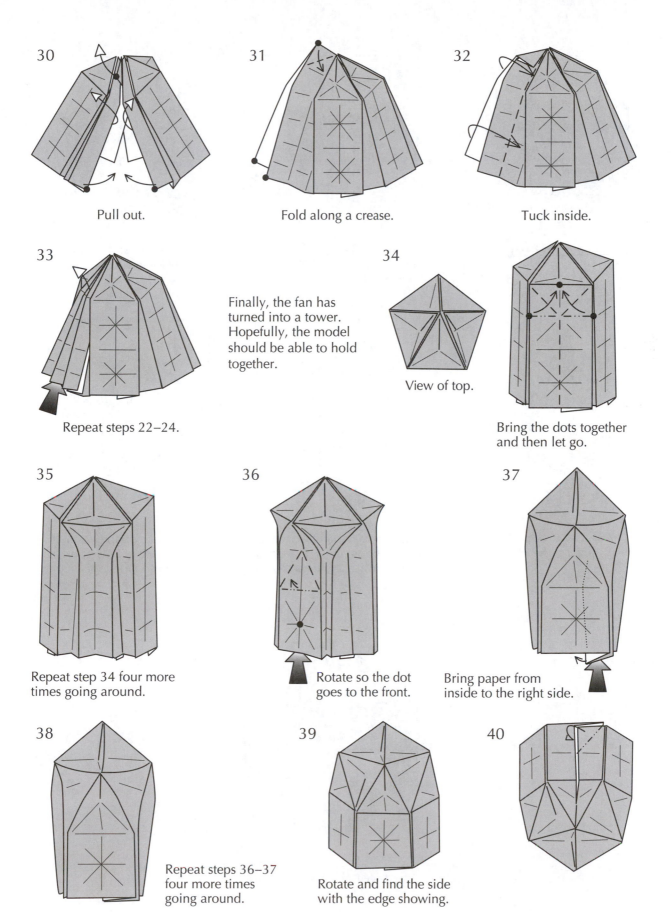

30

Pull out.

31

Fold along a crease.

32

Tuck inside.

33

Repeat steps 22–24.

Finally, the fan has turned into a tower. Hopefully, the model should be able to hold together.

34

View of top.

Bring the dots together and then let go.

35

Repeat step 34 four more times going around.

36

Rotate so the dot goes to the front.

37

Bring paper from inside to the right side.

38

Repeat steps 36–37 four more times going around.

39

Rotate and find the side with the edge showing.

40

41

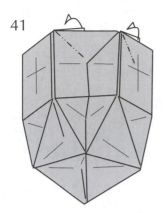

Fold all the loose edges
going all around.

42

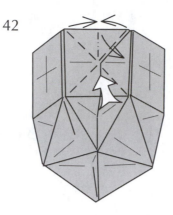

Form a sunken triangle
bringing the extra
paper to the front.

43

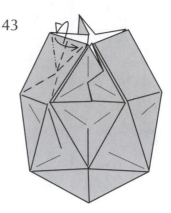

Form a sunken triangle
while folding the extra
paper inside.

44

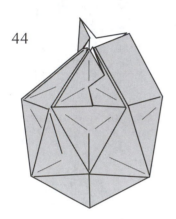

Repeat step 43
on the right.

45

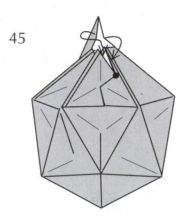

Form the last two triangles on the
back while bringing the extra paper
to the front inside the front triangle.

46

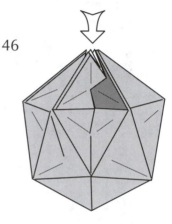

Bring the dark paper to the front
to lock the folds. Also push
down to keep the folds in place.

47

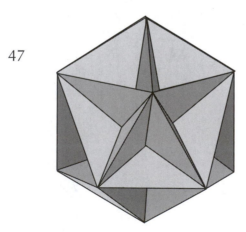

Sunken Icosahedron

Dipyramids and
Dimpled Dipyramids

Dipyramid Design

A pyramid has a polygon base and triangular sides whose vertices meet at the top. Dipyramids, or diamonds, are composed of a pair of identical pyramids joined at the base. The sides are identical isosceles triangles. For a polygon with n sides, there are $2n$ triangles in a dipyramid. A dipyramid is defined by its polygon n and angle α at the top of the triangle. The variety of dipyramids is rich.

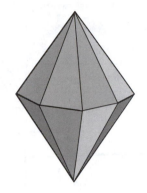
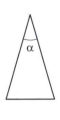

Height/Diameter Formula

A measure of a dipyramid is the proportion of the height to its diameter. One way to calculate that is to find the ratio d/c, where d is the distance from the center of the polyhedron to the top vertex (i.e., half the height) and c is the distance from the center to a vertex on the equator (i.e., half the diameter):

Let $H = d/c$

Assume that the length of each side of the polygon base is 1. Length b of the triangle can be found by looking at the right triangle created by splitting the face in half:

$$\sin(\alpha/2) = (1/2)/b$$

$$b = \frac{1}{2\sin(\alpha/2)}$$

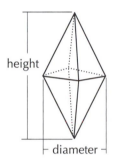

Length c on the polygon base is also one side of an isosceles triangle and can be found by the same process:

$$\sin((360°/n)/2) = (1/2)/c$$

$$c = \frac{1}{2\sin(180°/n)}$$

With these two measurements, angle γ is found:

$$\cos(\gamma) = c/b$$

$$\gamma = \arccos(c/b)$$

$$\tan(\gamma) = d/c = H$$

So given polygon n and angle α, the proportion of the dipyramid's height to its diameter is

$$H = \tan\left(\arccos\left(\frac{\sin(\alpha/2)}{\sin(180°/n))}\right)\right)$$

If you want to find α given H and n, then

$$\alpha = 2\arcsin[\sin(180°/n)\cos(\arctan(H))]$$

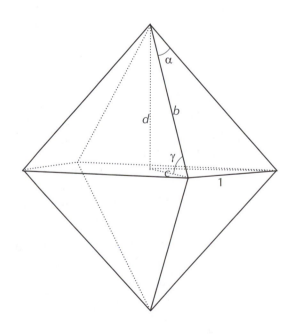

Groups of Dipyramids

1. Dipyramids inscibed in a sphere.

All of the vertices would rest on the surface of a sphere and so $H = 1$.

Since $\cos(\arctan(1)) = \cos 45° = 1/\sqrt{2}$

$$\alpha = 2 \arcsin[\sin(180°/n)/\sqrt{2}]$$

For a given polygon n, here are the angles α:

n	α
3	75.52°
4	60°
5	49.12°
6	41.41°
7	35.73°
8	31.40°
9	28.00°
10	25.24°
11	22.98°
12	21.09°

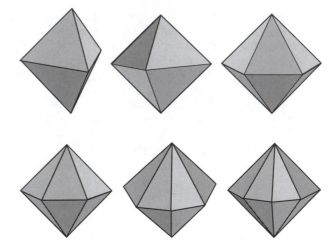

Dipyramids inscribed in a sphere from triangular to octagonal.

2. Duals of uniform prisms (prisms with two regular n-sided bases and n square sides).

$$\alpha = \arccos[1 - 2(\sin(180°/n))^4]$$

I thank Peter Messer for this formula, found in his paper "Closed-Form Expressions for Uniform Polyhedra and Their Duals" (*Discrete & Computational Geometry* 27:3 (2002), 353–375).

The uniform pentagonal prism and pentagonal dipyramid are dual pairs.

n	α	H
3	97.18°	.577350
4	60°	1
5	40.42°	1.37638
6	28.96°	1.73205
7	21.70°	2.07652
8	16.84°	2.41421
9	13.44°	2.74748
10	10.96°	3.07768
11	9.11°	3.40569
12	7.68°	3.73205

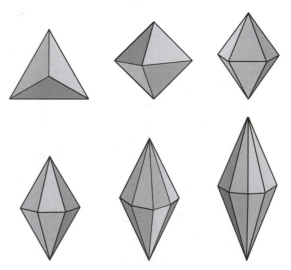

Dipyramids whose duals are uniform prisms from triangular to octagonal.

3. Dipyramids of the same polygon n. For example, for a square base (n = 4),

H	α
$1/\sqrt{2}$	70.53°
1	60°
$\sqrt{2}$	48.19°
2	36.87°

Squat Square Diamond

Octahedron

4. Dipyramids with the same H value (H ≠ 1). Varying the polygonal base will change the value of angle α, as in the first group.

5. Dipyramids with the same angle α while varying the polygonal base.

Silver Square Dipyramid

Tall Square Diamond

Layouts and Crease Patterns of Dipyramids

Triangular base.
(n = 3)

 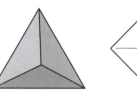

Even/odd symmetry

Even/odd symmetry

Square symmetry

Square base.
(n = 4)

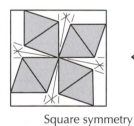 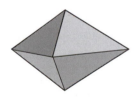

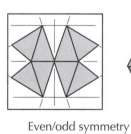 

Square symmetry

Even/odd symmetry

Base with odd number of sides.
(n = 5, 7, ...)

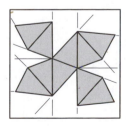

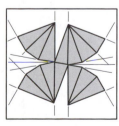 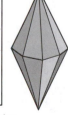

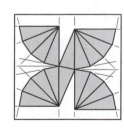 

Odd symmetry

Base with even number of sides.
(n = 6, 8, ...)

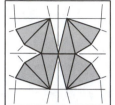  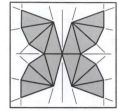  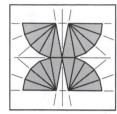

Even/odd symmetry

Landmarks

For dipyramids of polygons with an even number of sides (n = 4, 6, 8, ...), the layouts have even and odd symmetry. Two lines going through the center of the layout cross at angle α. Two horizontal lines distance b from the top and bottom edges determine the height of the faces. This value b defines the angle of the faces, α, and sets the layout.

Given α

$$\tan(\alpha/2) = (1/2 - a)/(1/2)$$
$$a = (1/2)(1 - \tan(\alpha/2))$$

The value b should be set as small as possible (to increase the size of the model) but large enough to allow for a tab; b can be chosen as an easy-to-find landmark.

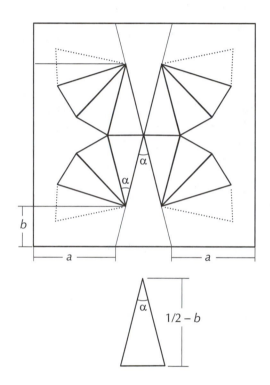

For dipyramids of polygons with an odd number of sides (n = 5, 7, ...), the layouts have odd symmetry. One line going through the center at an angle is related to the angles of the triangular faces. Two vertical lines distance b from the left and right edges are close to the center. This value b defines the angles of the faces and sets the layout.

Given α

$$\beta = (180° - \alpha)/2 = 90° - \alpha/2$$
$$\gamma = 90° - \beta = \alpha/2$$
$$\tan(\gamma) = \tan(\alpha/2) = 1 - 2a$$
$$a = (1/2)(1 - \tan(\alpha/2))$$

The value of b is between a and 1/2; it is chosen so that c is small but large enough to allow for a tab; b can be chosen as an easy-to-find landmark.

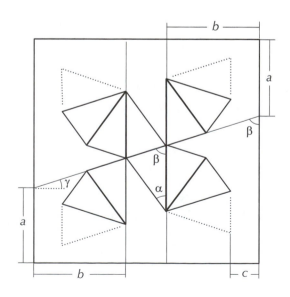

Using the layouts above, another interesting group of dipyramids would be those with crease patterns that include the horizontal line.

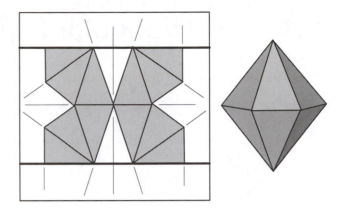

Find α give a polygon *n*,

$$(n/2)\alpha + \beta = 180°$$
$$\beta = (180° - \alpha)/2 = 90° - \alpha/2$$

so

$$(n/2)\alpha + 90° - \alpha/2 = 180°$$

$\alpha = 180°/(n-1)$

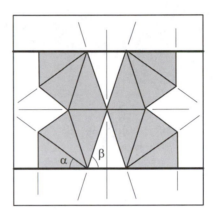

n	α
3	90°
4	60°
5	45°
6	36°
7	30°
8	25.714°
9	22.5°
10	20°
11	18°
12	16.3636°

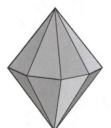

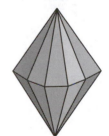

From triangular to decagonal dipyramids.

Triangular Dipyramid 90°

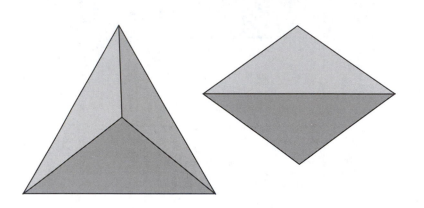
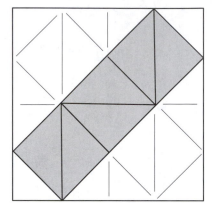

The angles of each of the six triangles are 90°, 45°, and 45°.

1

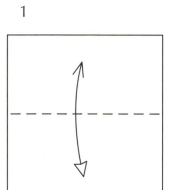

Fold and unfold.

2

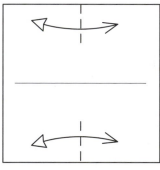

Fold and unfold at
the top and bottom.

3

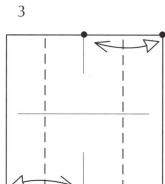

Fold and unfold.

4

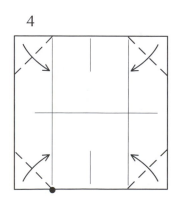

5

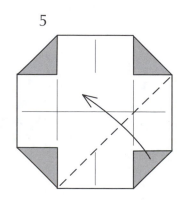

6

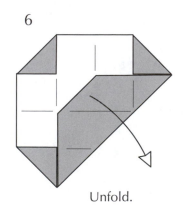

Unfold.

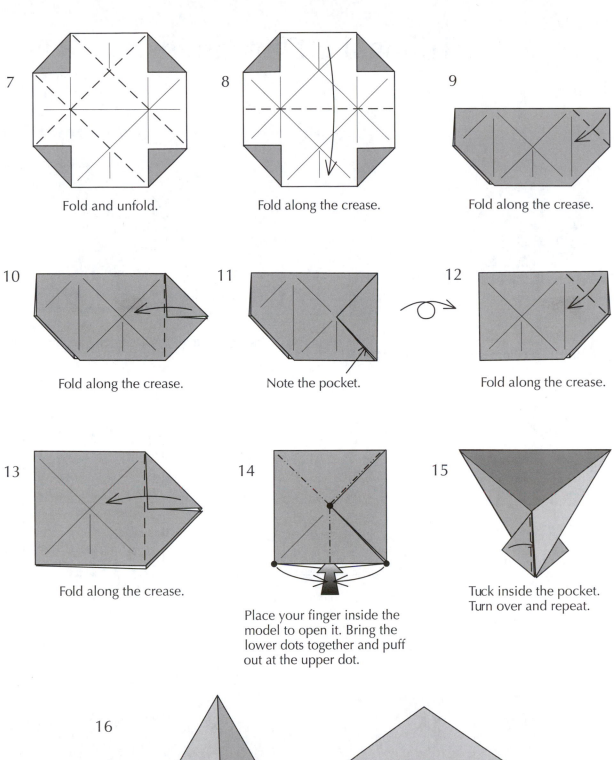

7 Fold and unfold.

8 Fold along the crease.

9 Fold along the crease.

10 Fold along the crease.

11 Note the pocket.

12 Fold along the crease.

13 Fold along the crease.

14 Place your finger inside the model to open it. Bring the lower dots together and puff out at the upper dot.

15 Tuck inside the pocket. Turn over and repeat.

16

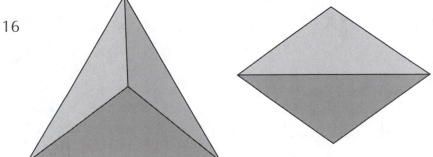

Triangular Dipyramid 90°

Triangular Dipyramid

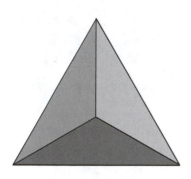

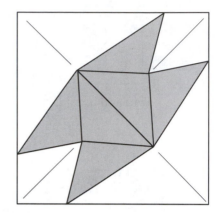

This dipyramid is the dual of the triangular uniform prism. Each of the six triangles has sides with proportions 2, 2, and 3. The angles are 41.41°, 41.41°, and 97.18°.

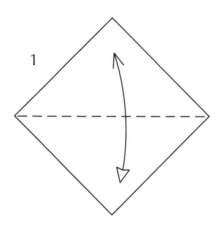

1

Fold and unfold. Rotate.

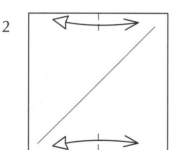

2

Fold and unfold at
the top and bottom.

3

Bring the left edge to the lower
dot. Crease at the intersection.

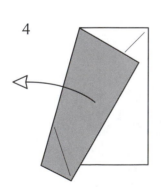

4

Unfold and rotate 180°.

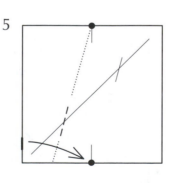

5

Repeat steps 3–4.
Turn over and rotate.

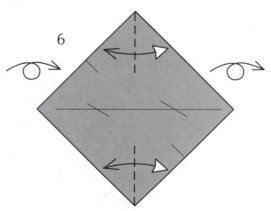

6

Fold and unfold but do
not crease in the center.

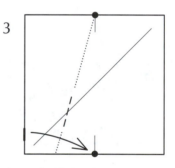

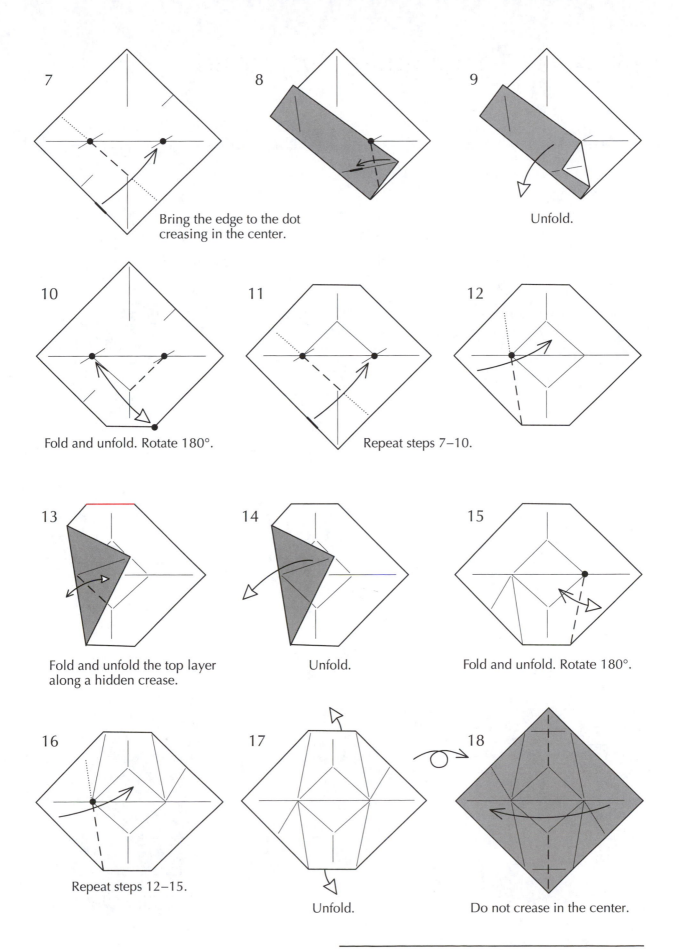

7

Bring the edge to the dot creasing in the center.

8

9

Unfold.

10

Fold and unfold. Rotate 180°.

11

Repeat steps 7–10.

12

13

Fold and unfold the top layer along a hidden crease.

14

Unfold.

15

Fold and unfold. Rotate 180°.

16

Repeat steps 12–15.

17

Unfold.

18

Do not crease in the center.

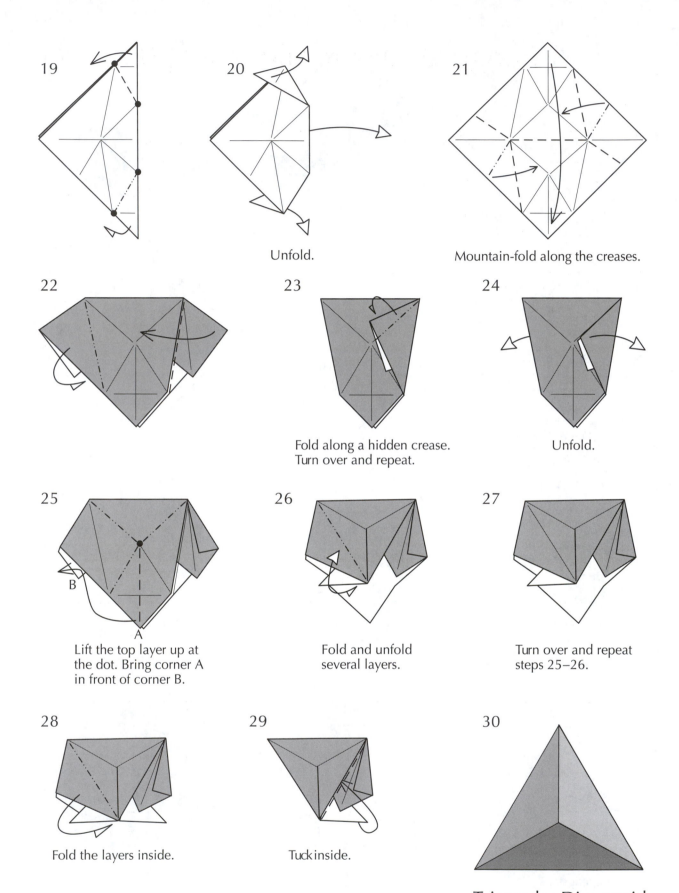

19

20

Unfold.

21

Mountain-fold along the creases.

22

23

Fold along a hidden crease.
Turn over and repeat.

24

Unfold.

25

B

A

Lift the top layer up at
the dot. Bring corner A
in front of corner B.

26

Fold and unfold
several layers.

27

Turn over and repeat
steps 25–26.

28

Fold the layers inside.

29

Tuck inside.

30

Triangular Dipyramid

Triangular Dipyramid in a Sphere

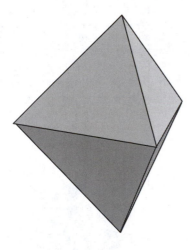

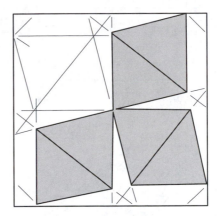

This triangular dipyramid is inscibed in a sphere. The crease pattern shows 3/4 square symmetry. The angles of each of the six triangles are 75.52°, 52.24°, and 52.24°.

1

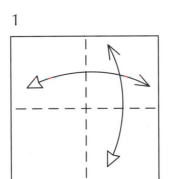

Fold and unfold.

2

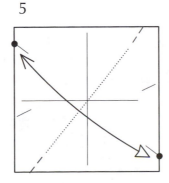

Fold and unfold on the left.

3

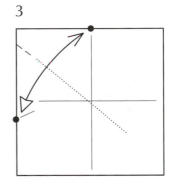

Fold and unfold on the left. Rotate 180°.

4

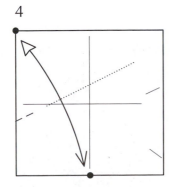

Repeat steps 2–3.

5

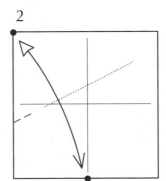

Fold and unfold creasing at the edges.

6

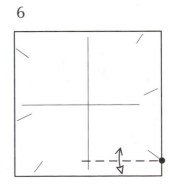

Fold and unfold.

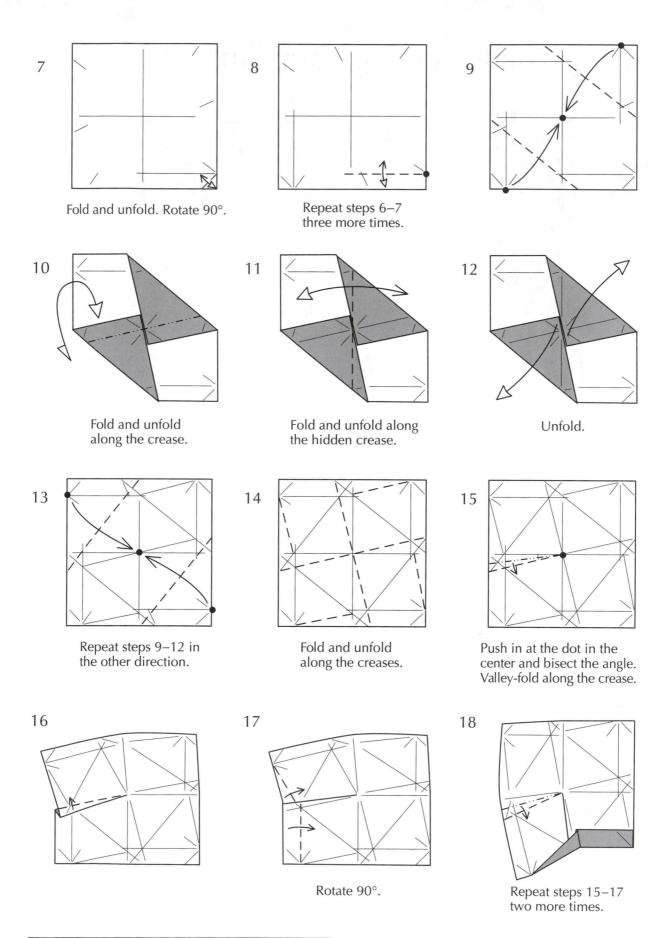

7

Fold and unfold. Rotate 90°.

8

Repeat steps 6–7
three more times.

9

10

Fold and unfold
along the crease.

11

Fold and unfold along
the hidden crease.

12

Unfold.

13

Repeat steps 9–12 in
the other direction.

14

Fold and unfold
along the creases.

15

Push in at the dot in the
center and bisect the angle.
Valley-fold along the crease.

16

17

Rotate 90°.

18

Repeat steps 15–17
two more times.

19

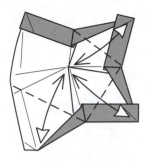

Fold and unfold on three of the four sides.

20

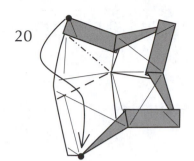

Valley-fold along the crease.

21

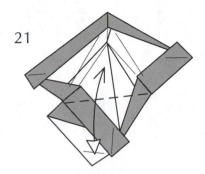

Fold and unfold all the layers.

22

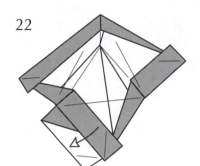

Unfold.

23

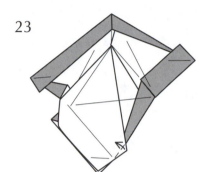

24

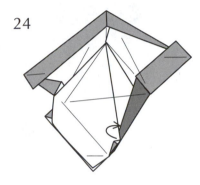

Tuck under the darker layer.

25

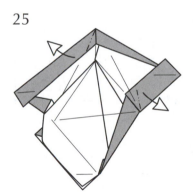

Lift up the two flaps.

26

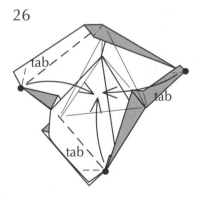

Bring the three dots together and close the model with a three-way twist lock.

27

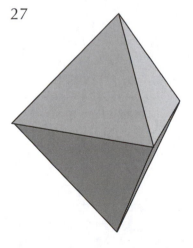

Triangular Dipyramid in a Sphere

Tall Triangular Dipyramid

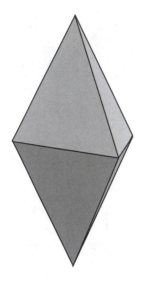

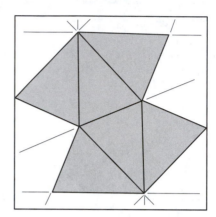

This is one of the simplest dipyramids in this section, but it uses several of the same folding techniques as other ones. The angles of each of the six triangles are 45°, 37.5°, and 37.5°.

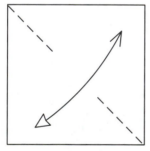

1

Fold and unfold
creasing at the ends.

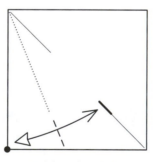

2

Fold and unfold
at the bottom.

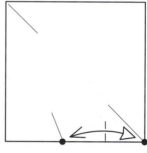

3

Fold and unfold.
Rotate 180°.

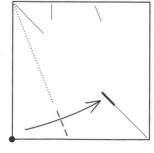

4

Repeat steps 2–3.

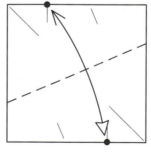

5

Fold and unfold.

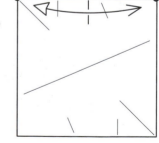

6

Fold and unfold.

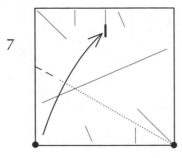

7

Bring the lower left corner to the crease. Fold on the left.

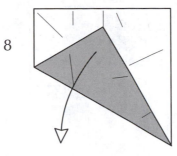

8

Unfold.

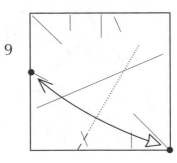

9

Fold and unfold at the bottom.

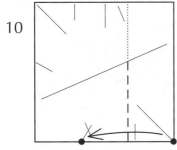

10

Crease under the center line.

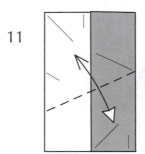

11

Fold and unfold along a partially hidden crease.

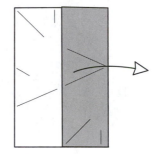

12

Unfold and rotate 180°.

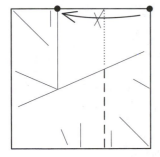

13

Repeat steps 10–12.

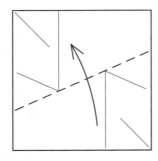

14

Fold along the crease.

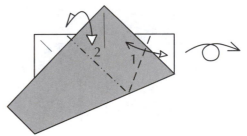

15

Fold and unfold along the creases.

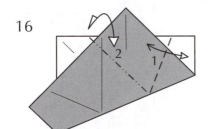

16

Fold and unfold.

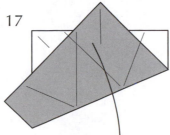

17

Unfold.

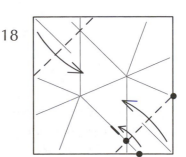

18

19

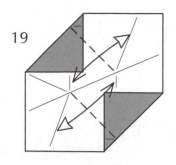

Fold and unfold
along the creases.

20

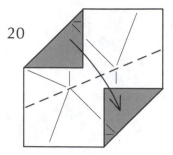

Rotate.

21

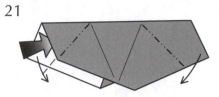

Reverse folds.

22

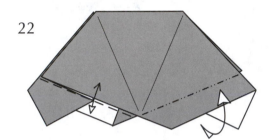

Fold and unfold.

23

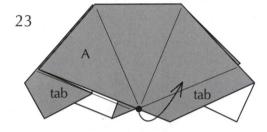

Lift up at the dot and the model will
become 3D. Rotate to view the opening
so that region A is at the top right.

24

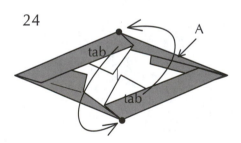

Tuck and interlock the
tabs. The dots will meet.

25

Tall Triangular Dipyramid

Tall Square Dipyramid

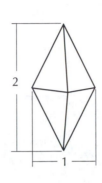

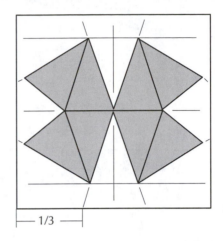

The height of this square dipyramid is twice the diameter. According to the formulas for H and α derived earlier, the small angle in each triangle is about 36.87° and a convenient landmark of 1/3 is used to achieve the dimensions.

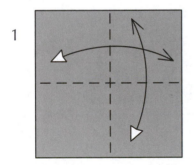

1 Fold and unfold.

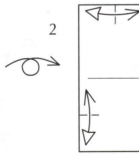

2 Fold and unfold to find the quarter marks.

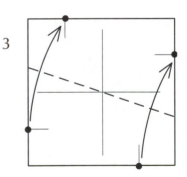

3

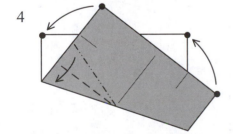

4 Valley-fold along the crease. Turn over and repeat.

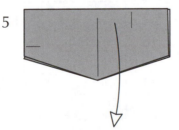

5 Unfold and rotate 90°.

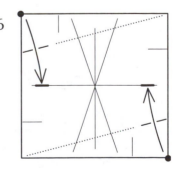

6 Crease at the edges.

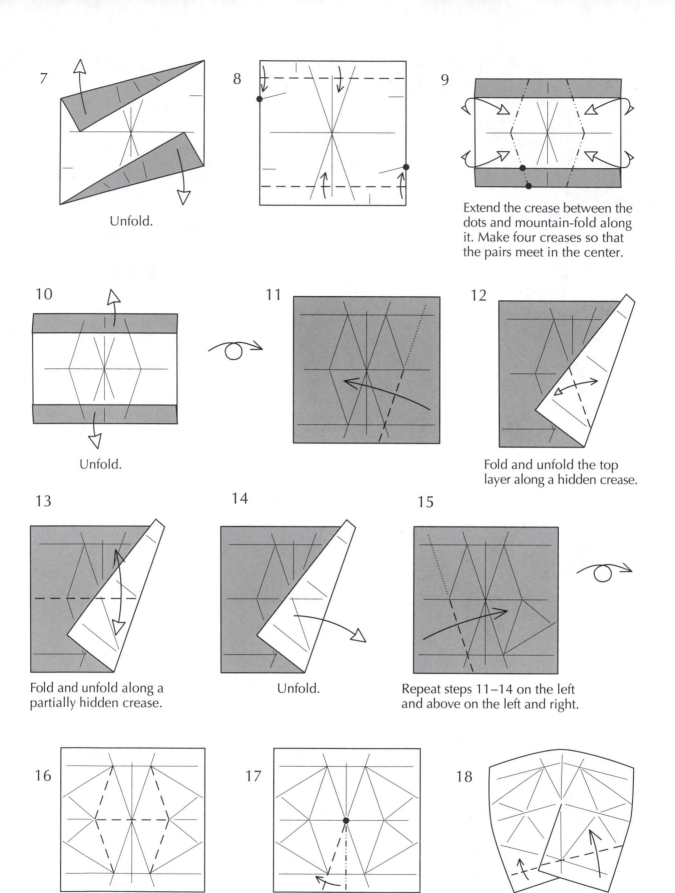

7

Unfold.

8

9

Extend the crease between the dots and mountain-fold along it. Make four creases so that the pairs meet in the center.

10

Unfold.

11

12

Fold and unfold the top layer along a hidden crease.

13

Fold and unfold along a partially hidden crease.

14

Unfold.

15

Repeat steps 11–14 on the left and above on the left and right.

16

Fold and unfold along the creases.

17

Push in at the dot.

18

Fold along the creases.

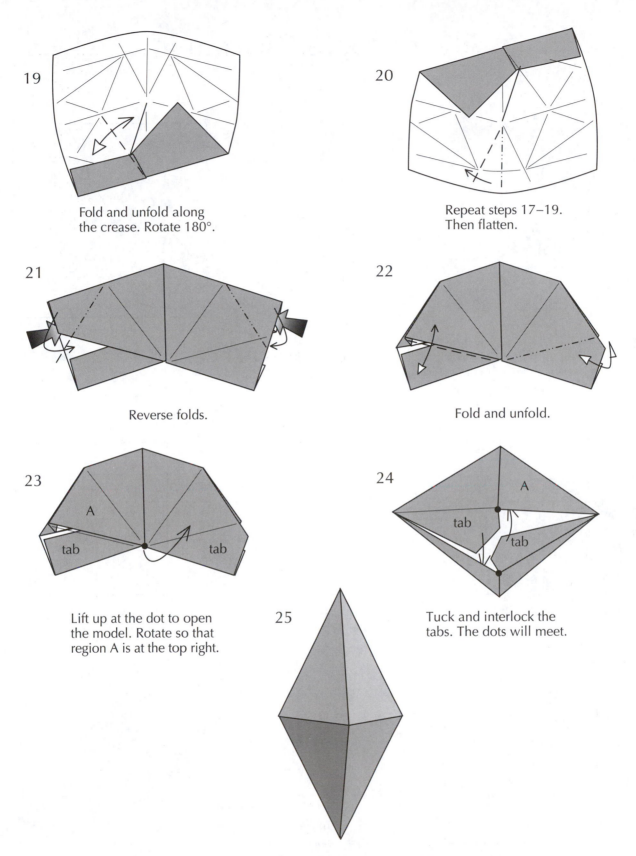

19

Fold and unfold along
the crease. Rotate 180°.

20

Repeat steps 17–19.
Then flatten.

21

Reverse folds.

22

Fold and unfold.

23

A

tab   tab

Lift up at the dot to open
the model. Rotate so that
region A is at the top right.

24

A

tab

tab

Tuck and interlock the
tabs. The dots will meet.

25

Tall Square Dipyramid

Silver Square Dipyramid

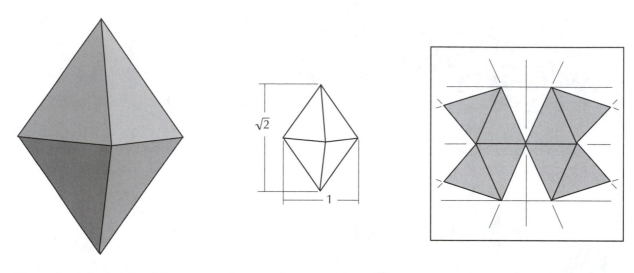

The ratio of the height of this square dipyramid to its width is $\sqrt{2}$ to 1. The small angle in each triangle is about 48° to achieve the dimensions.

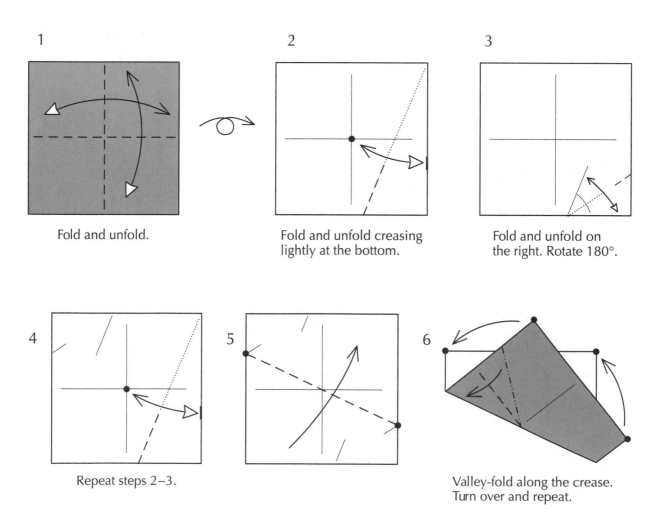

1
Fold and unfold.

2
Fold and unfold creasing lightly at the bottom.

3
Fold and unfold on the right. Rotate 180°.

4
Repeat steps 2–3.

5

6
Valley-fold along the crease. Turn over and repeat.

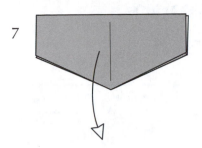

7

Unfold and rotate 90°.

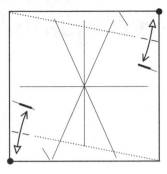

8

Fold and unfold
on the edges.

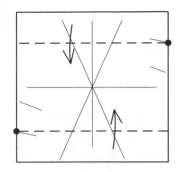

9

Continue with steps 11–20
of the Tall Square Dipyramid.

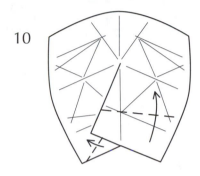

10

Fold along the crease on
the right and rotate 180°.

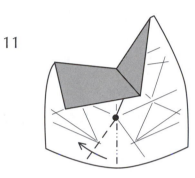

11

Push in at the dot. Repeat step 10 and
flatten. Then continue with step 24 through
the end of the Tall Square Dipyramid.

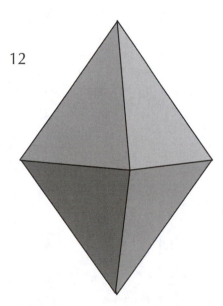

12

Silver Square Dipyramid

Squat Silver Square Diamond

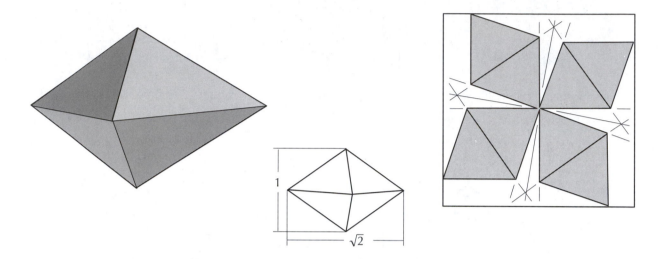

The ratio of the height of this square dipyramid to its diameter is 1 to $\sqrt{2}$. The angles of each of the eight triangles are 70.53°, 54.74°, and 54.74°. The crease pattern shows square symmetry.

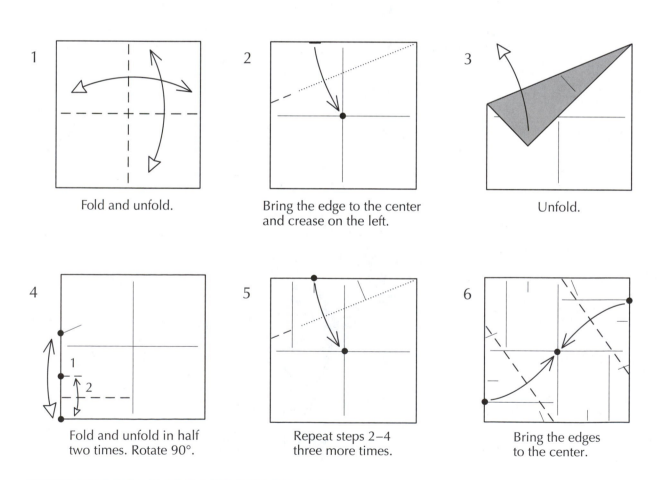

1 Fold and unfold.

2 Bring the edge to the center and crease on the left.

3 Unfold.

4 Fold and unfold in half two times. Rotate 90°.

5 Repeat steps 2–4 three more times.

6 Bring the edges to the center.

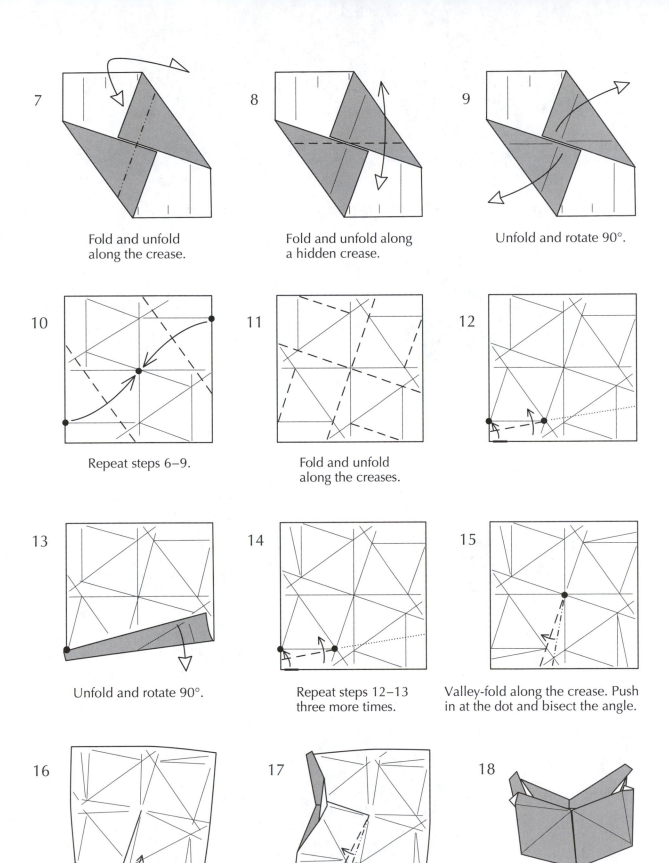

7 — Fold and unfold along the crease.

8 — Fold and unfold along a hidden crease.

9 — Unfold and rotate 90°.

10 — Repeat steps 6–9.

11 — Fold and unfold along the creases.

12

13 — Unfold and rotate 90°.

14 — Repeat steps 12–13 three more times.

15 — Valley-fold along the crease. Push in at the dot and bisect the angle.

16 — Flatten to form a triangle. Rotate 90°.

17 — Repeat steps 15–16 three more times. Rotate to view the outside.

18 — Flatten.

19

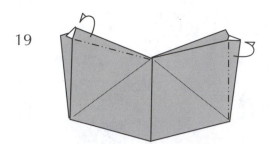

Turn over and repeat.

20

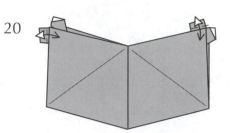

Fold and unfold. Turn over and repeat.

21

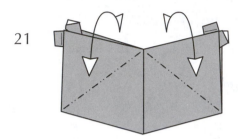

Open to fold inside and unfold. Repeat behind.

22

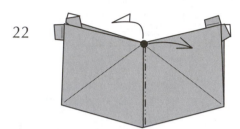

Open and flatten. Follow the dot in the next step.

23

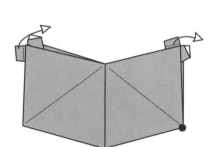

Unfold the thin flaps. Turn over and repeat.

24

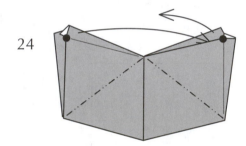

The model becomes 3D as the four dots are brought together. Close the model by interlocking the four tabs. These tabs spiral inward with a twist lock.

25

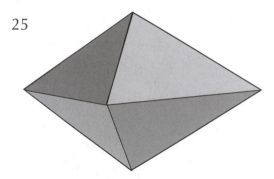

Squat Silver Square Dipyramid

Pentagonal Dipyramid

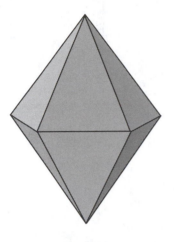

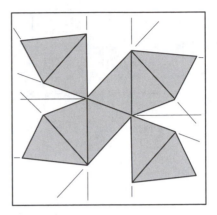

This ten-sided dipyramid is the dual of the uniform pentagonal prism. The small angle in each triangle is about 41°.

1

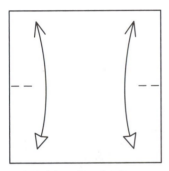

Fold and unfold on the left and right.

2

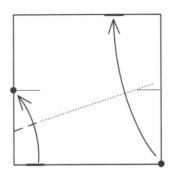

Bring the lower right corner to the top edge and the bottom edge to the left center. Crease on the left.

3

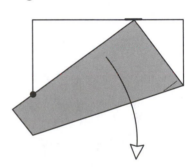

Unfold.

4

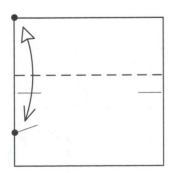

Fold and unfold. Rotate 180°.

5

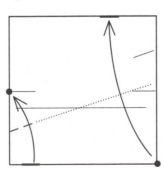

Repeat steps 2–4. Rotate 90°.

6

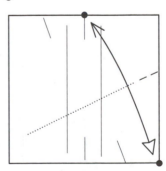

Fold and unfold.

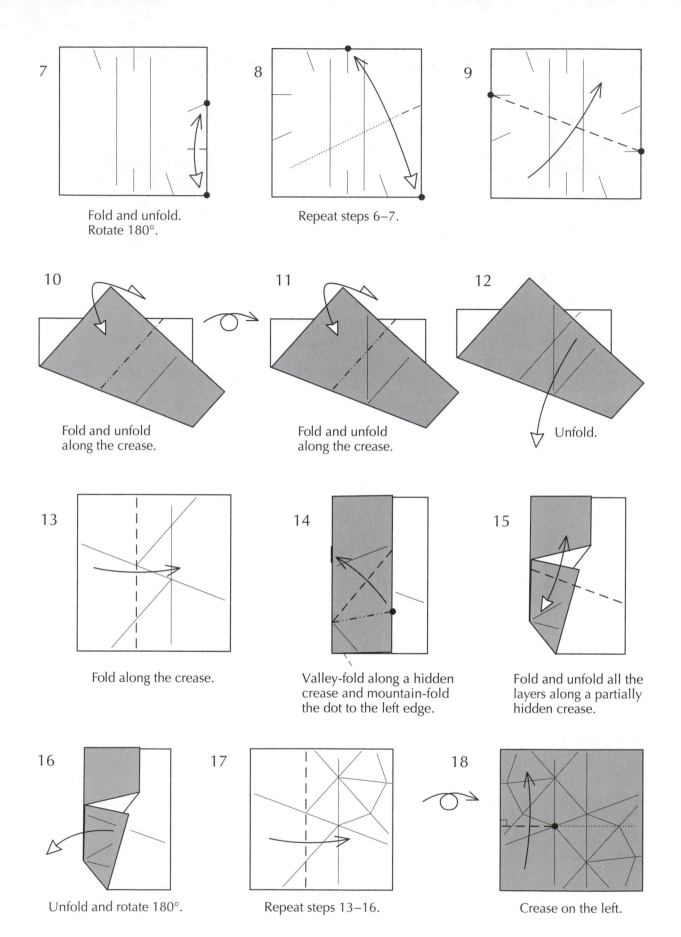

7

Fold and unfold.
Rotate 180°.

8

Repeat steps 6–7.

9

10

Fold and unfold
along the crease.

11

Fold and unfold
along the crease.

12

Unfold.

13

Fold along the crease.

14

Valley-fold along a hidden
crease and mountain-fold
the dot to the left edge.

15

Fold and unfold all the
layers along a partially
hidden crease.

16

Unfold and rotate 180°.

17

Repeat steps 13–16.

18

Crease on the left.

19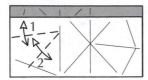

Fold and unfold the top layer along hidden creases.

20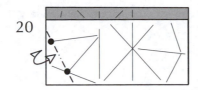

Fold and unfold both layers along the hidden crease between the dots.

21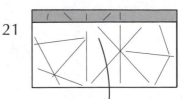

Unfold and rotate 180°.

22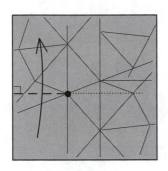

Repeat steps 18–21.

23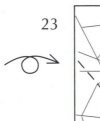

Fold and unfold along the creases.

24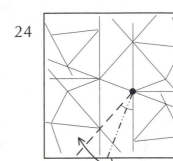

Push in at the dot and bisect the angle.

25

On the right, fold along the crease between the dots. On the left, bring the edge to the dot.

26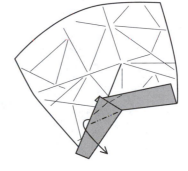

Open while folding down. Then flatten to form a triangle. Rotate 180°.

27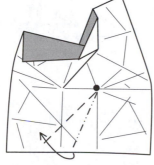

Repeat steps 24–26. Then flatten.

28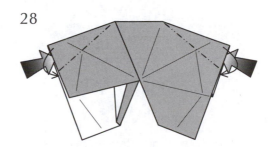

Reverse folds.

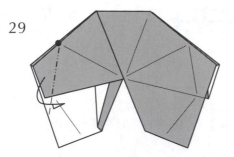

29

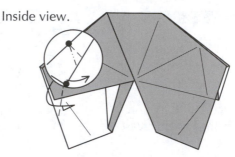

Inside view.

Fold the inside layers together by this method: Hold at the upper dot and pull the inside flap at the lower dot toward the center. The model locks at the upper dot. This fold is used in other polyhedra, especially in more dipyramids. This will be called the spine-lock fold. Turn over and repeat.

30

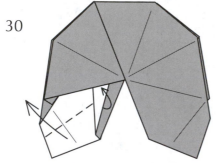

Refold as it was in step 26. Turn over and repeat.

31

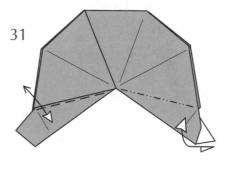

Fold and unfold.

32

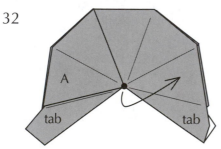

Lift up at the dot and the model will become 3D. Rotate to view the opening so that region A is at the top right.

33

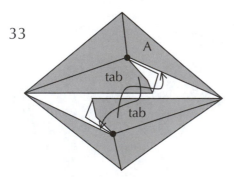

Tuck and interlock the tabs. The dots will meet.

34

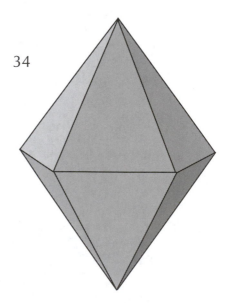

Pentagonal Dipyramid

Pentagonal Dipyramid 45°

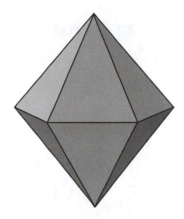
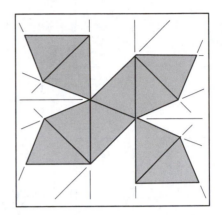

This ten-sided dipyramid is similar in folding to the Pentagonal Dipyramid. The crease pattern shows an easier layout since the small angle in each triangle is 45°.

1
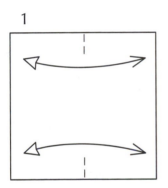

Fold and unfold on the top and bottom.

2
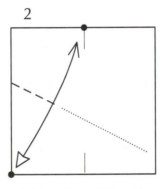

Fold and unfold on the left creasing lightly.

3
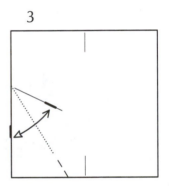

Fold and unfold at the bottom.

4
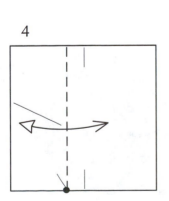

Fold and unfold. Rotate 180°.

5
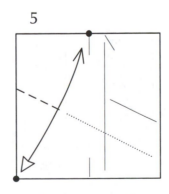

Repeat steps 2–4.

6
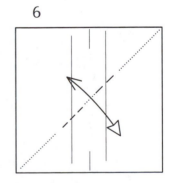

Fold and unfold creasing lightly.

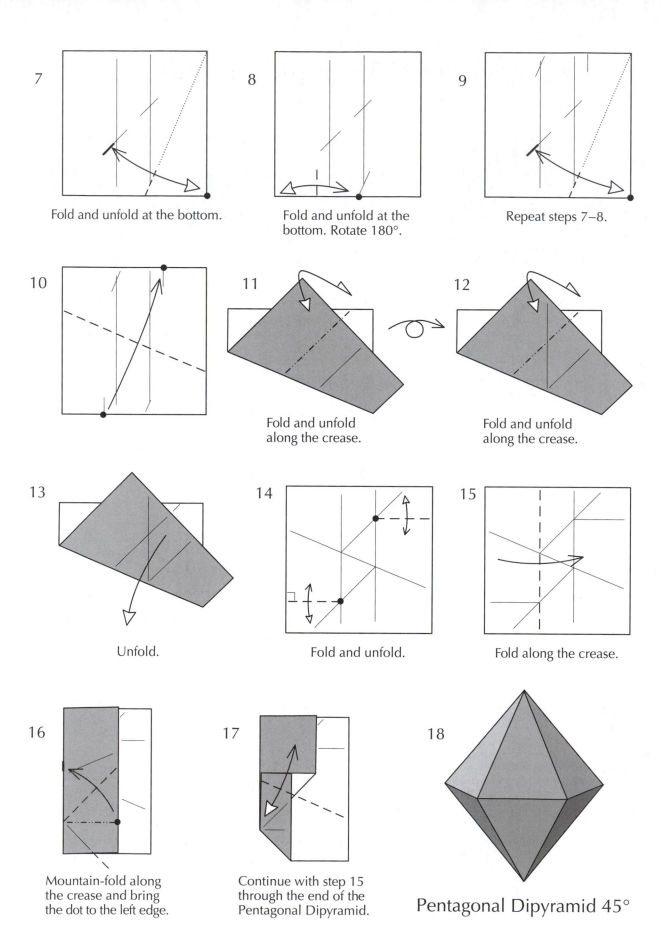

7

Fold and unfold at the bottom.

8

Fold and unfold at the bottom. Rotate 180°.

9

Repeat steps 7–8.

10

11

Fold and unfold along the crease.

12

Fold and unfold along the crease.

13

Unfold.

14

Fold and unfold.

15

Fold along the crease.

16

Mountain-fold along the crease and bring the dot to the left edge.

17

Continue with step 15 through the end of the Pentagonal Dipyramid.

18

Pentagonal Dipyramid 45°

Pentagonal Dipyramid in a Sphere

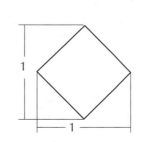

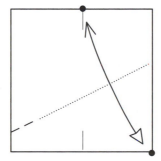

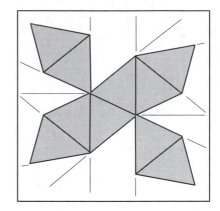

This is the pentagonal dipyramid inscibed in a sphere. The ratio of the height to the diameter is 1 to 1. The angles of each of the ten triangles are 49.12°, 65.44°, and 65.44°.

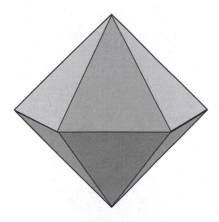

1

Fold and unfold on
the top and bottom.

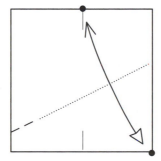

2

Fold and unfold on the left.

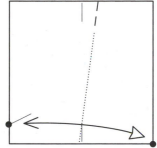

3

Fold and unfold at the top.

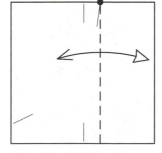

4

Fold and unfold. Rotate 180°.

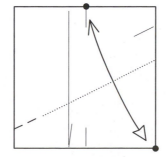

5

Repeat steps 2–4. Rotate 90°.

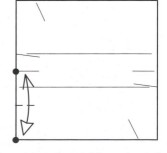

6

Fold and unfold on the left.

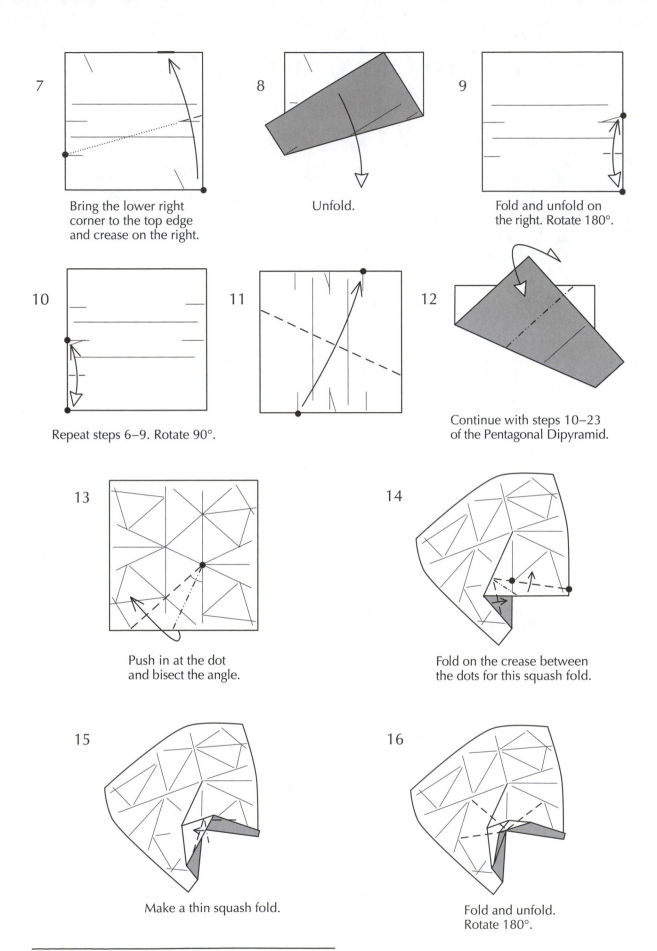

7 Bring the lower right corner to the top edge and crease on the right.

8 Unfold.

9 Fold and unfold on the right. Rotate 180°.

10 Repeat steps 6–9. Rotate 90°.

11

12 Continue with steps 10–23 of the Pentagonal Dipyramid.

13 Push in at the dot and bisect the angle.

14 Fold on the crease between the dots for this squash fold.

15 Make a thin squash fold.

16 Fold and unfold. Rotate 180°.

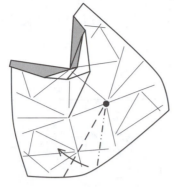

17

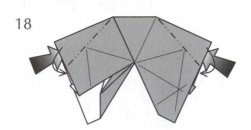

18

Repeat steps 13–16.
Then flatten.

Reverse folds.

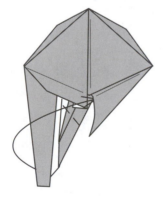

19

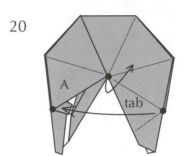

20

Fold the inside layers together for this
spine-lock fold. Turn over and repeat.

Lift up at the upper dot. The tab
goes under A as the dots meet.

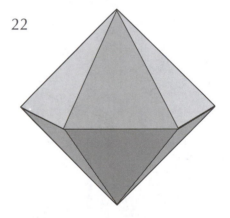

21

22

Tuck inside.

Pentagonal Dipyramid
in a Sphere

Golden Pentagonal Dipyramid

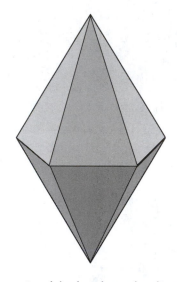 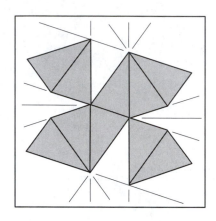

The ratio of the height to the diameter of this pentagonal dipyramid is 1.61803 to 1. The small angle of each triangle is 36°. The folding is similar to that of the Pentagonal Dipyramid.

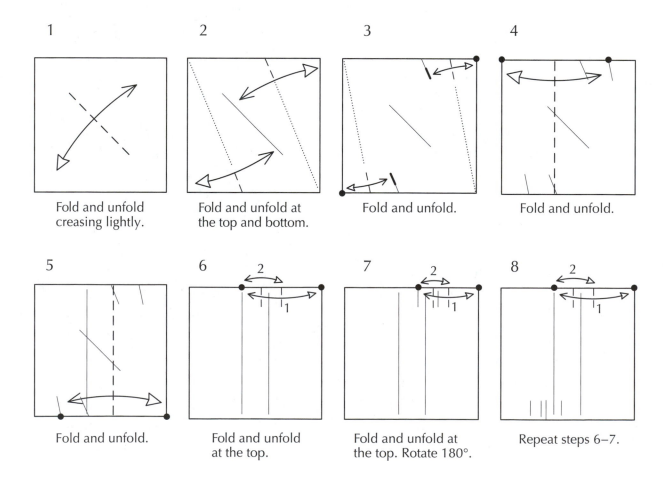

1

Fold and unfold creasing lightly.

2

Fold and unfold at the top and bottom.

3

Fold and unfold.

4

Fold and unfold.

5

Fold and unfold.

6

Fold and unfold at the top.

7

Fold and unfold at the top. Rotate 180°.

8

Repeat steps 6–7.

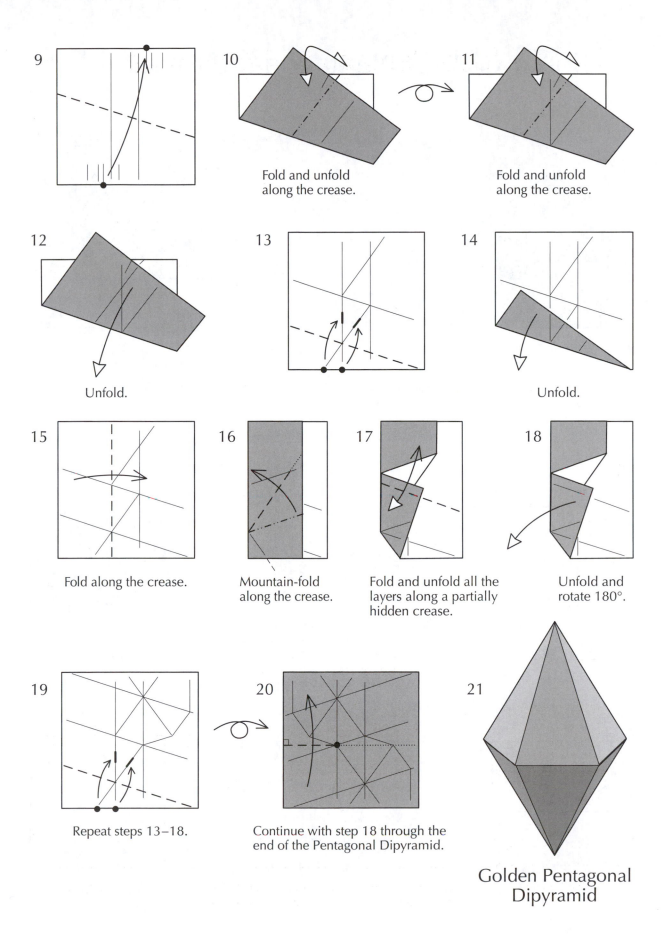

9

10 Fold and unfold
along the crease.

11 Fold and unfold
along the crease.

12 Unfold.

13

14 Unfold.

15 Fold along the crease.

16 Mountain-fold
along the crease.

17 Fold and unfold all the
layers along a partially
hidden crease.

18 Unfold and
rotate 180°.

19 Repeat steps 13–18.

20 Continue with step 18 through the
end of the Pentagonal Dipyramid.

21

Golden Pentagonal
Dipyramid

Squat Golden Pentagonal Dipyramid

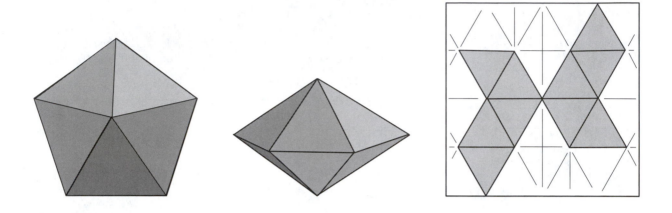

All the sides of this pentagonal dipyramid are equilateral triangles. The ratio of the height to the diameter is .61803 to 1.

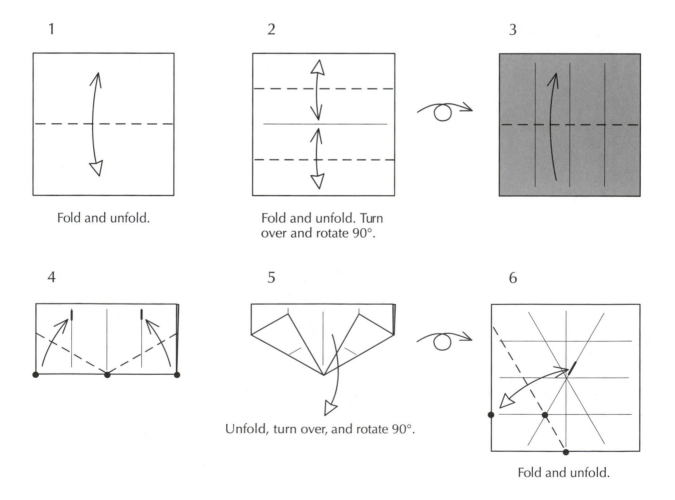

1

Fold and unfold.

2

Fold and unfold. Turn over and rotate 90°.

3

4

5

Unfold, turn over, and rotate 90°.

6

Fold and unfold.

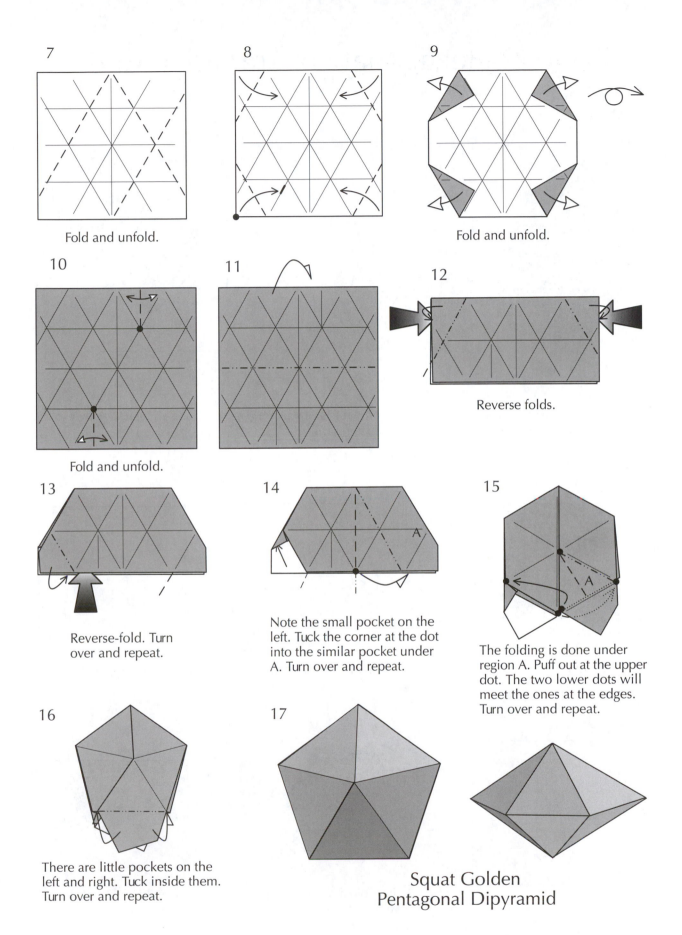

7

Fold and unfold.

8

9

Fold and unfold.

10

Fold and unfold.

11

12

Reverse folds.

13

Reverse-fold. Turn over and repeat.

14

Note the small pocket on the left. Tuck the corner at the dot into the similar pocket under A. Turn over and repeat.

15

The folding is done under region A. Puff out at the upper dot. The two lower dots will meet the ones at the edges. Turn over and repeat.

16

There are little pockets on the left and right. Tuck inside them. Turn over and repeat.

17

Squat Golden
Pentagonal Dipyramid

Hexagonal Dipyramid

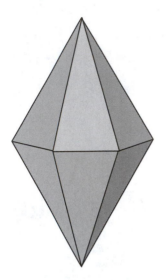

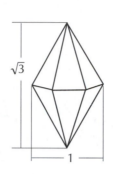

$\sqrt{3}$

1

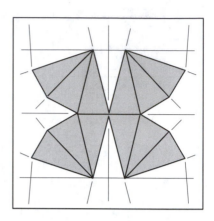

This 12-sided dipyramid is the dual of the uniform hexagonal prism. The small angle in each triangle is about 29°, and the ratio of its height to the diameter is $\sqrt{3}$ to 1.

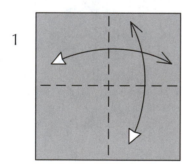

1

Fold and unfold.

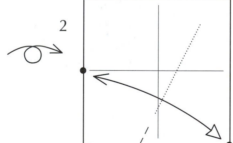

2

Fold and unfold at the bottom.

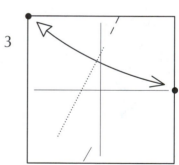

3

Fold and unfold at the top.

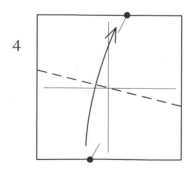

4

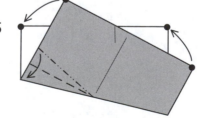

5

Valley-fold along the crease.
Turn over and repeat.

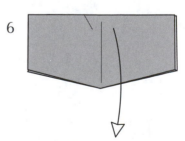

6

Unfold and rotate 90°.

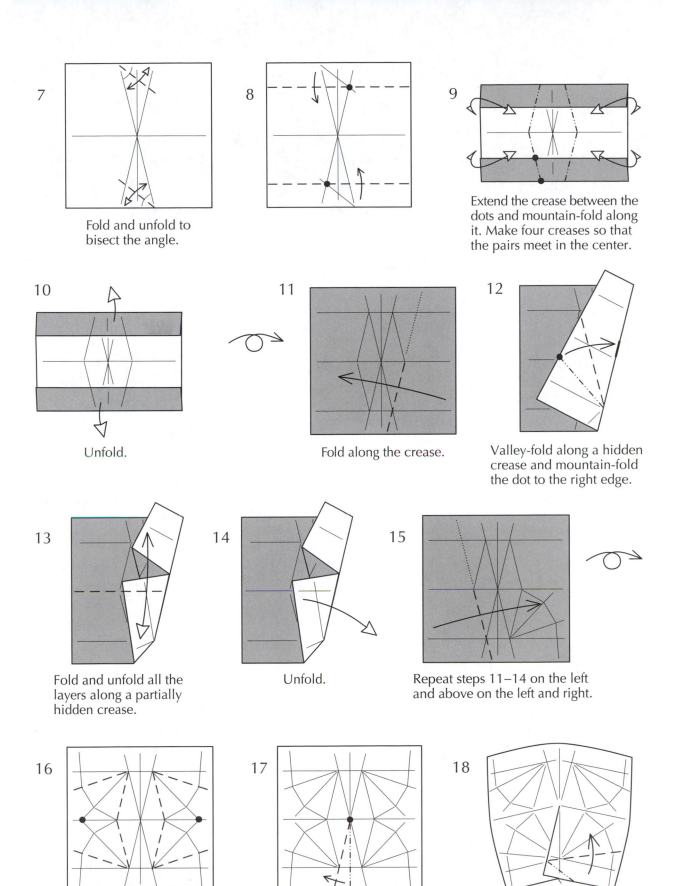

7

Fold and unfold to
bisect the angle.

8

9

Extend the crease between the
dots and mountain-fold along
it. Make four creases so that
the pairs meet in the center.

10

Unfold.

11

Fold along the crease.

12

Valley-fold along a hidden
crease and mountain-fold
the dot to the right edge.

13

Fold and unfold all the
layers along a partially
hidden crease.

14

Unfold.

15

Repeat steps 11–14 on the left
and above on the left and right.

16

Note the intersections at the dots.
Fold and unfold along the creases.

17

Push in at the dot.

18

Squash-fold and rotate 180°.

19

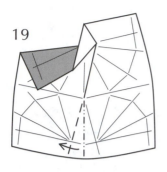

Repeat steps 17–18.
Then flatten.

20

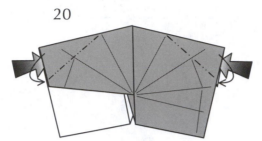

Reverse folds.

21

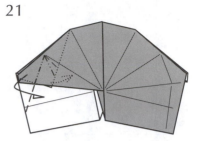

Fold the inside layers together
for this spine-lock fold. Turn
over and repeat.

22

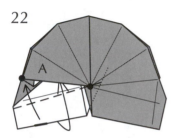

Bring the edge to the dot
and tuck under region A.
Turn over and repeat.

23

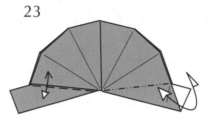

Fold and unfold.

24

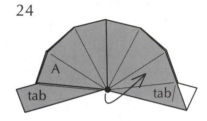

Lift up at the dot and the model
will become 3D. Rotate to view
the opening so that region A is
at the top right.

25

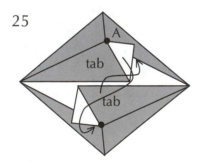

Tuck and interlock the
tabs. The dots will meet.

26

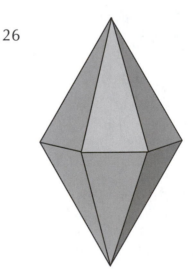

Hexagonal Dipyramid

Silver Hexagonal Dipyramid

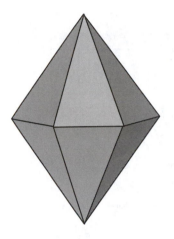

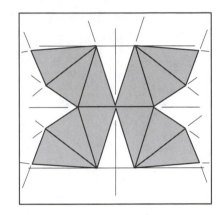

The ratio of the height of this dipyramid to its diameter is $\sqrt{2}$ to 1. The small angle in each triangle is about 34° to achieve the dimensions. The folding is similar to the Hexagonal Dipyramid.

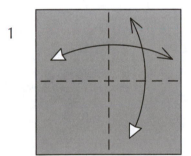

1

Fold and unfold.

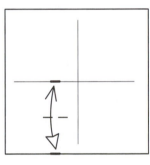

2

Fold and unfold creasing lightly.

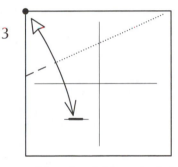

3

Fold and unfold on the left.

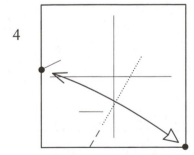

4

Fold and unfold at the bottom. Rotate 180°.

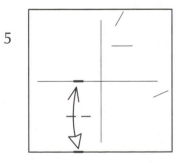

5

Repeat steps 2–4.

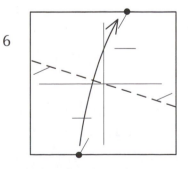

6

Continue with step 4 through the end of the Hexagonal Dipyramid.

Hexagonal Dipyramid 36°

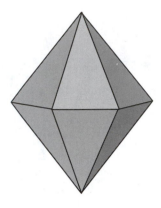

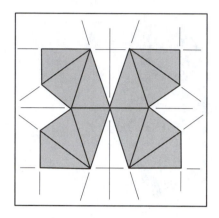

The geometry of this dipyramid makes it easier to fold since the horizontal creases are part of the crease pattern. This is achieved by setting the small angle of each triangle equal to 36°. The folding is similar to the Hexagonal Dipyramid.

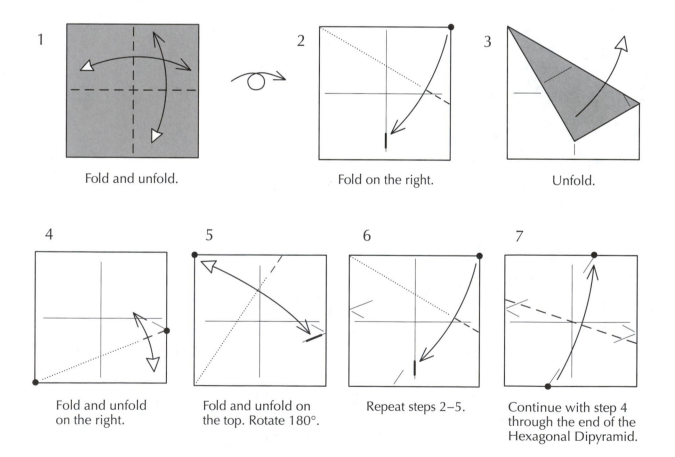

1. Fold and unfold.

2. Fold on the right.

3. Unfold.

4. Fold and unfold on the right.

5. Fold and unfold on the top. Rotate 180°.

6. Repeat steps 2–5.

7. Continue with step 4 through the end of the Hexagonal Dipyramid.

Hexagonal Dipyramid in a Sphere

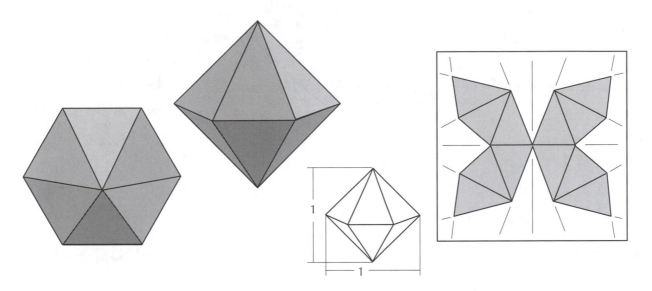

This dipyramid is inscibed in a sphere. The angles of each of the twelve triangles are 41.41°, 69.3°, and 69.3°, and the lengths of the sides are proportional to 1, $\sqrt{2}$, and $\sqrt{2}$.

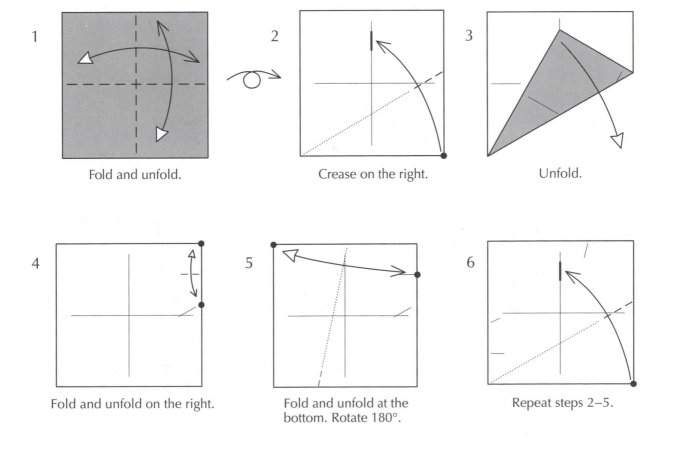

1 Fold and unfold.

2 Crease on the right.

3 Unfold.

4 Fold and unfold on the right.

5 Fold and unfold at the bottom. Rotate 180°.

6 Repeat steps 2–5.

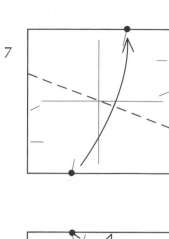

7

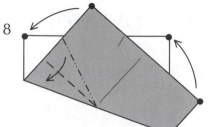

8

Valley-fold along the crease.
Turn over and repeat.

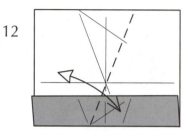

9

Unfold and rotate 90°.

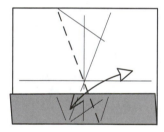

10

Fold and unfold to
bisect the angle.

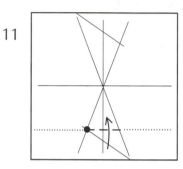

11

Crease in the center.

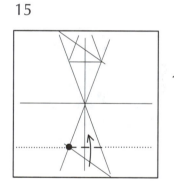

12

Fold and unfold.

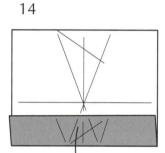

13

Fold and unfold.

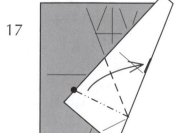

14

Unfold and rotate 180°.

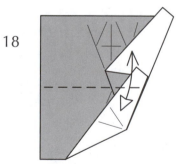

15

Repeat steps 11–14.

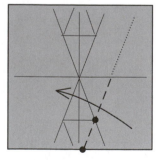

16

Extend the crease
along the dots.

17

Valley-fold along a hidden
crease and mountain-fold
the dot to the right edge.

18

Fold and unfold all the layers
along a partially hidden crease.

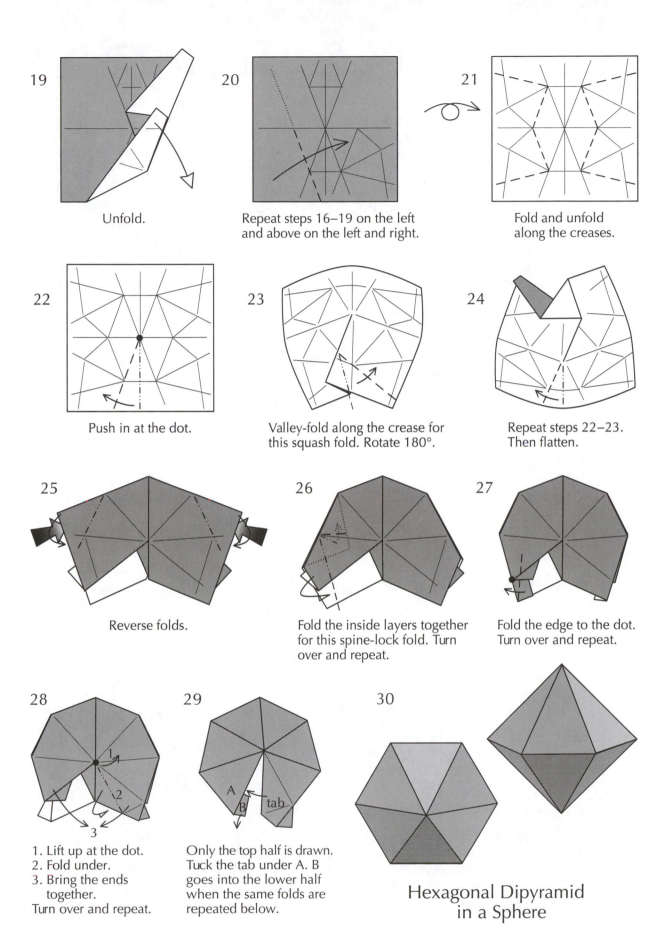

19

Unfold.

20

Repeat steps 16–19 on the left
and above on the left and right.

21

Fold and unfold
along the creases.

22

Push in at the dot.

23

Valley-fold along the crease for
this squash fold. Rotate 180°.

24

Repeat steps 22–23.
Then flatten.

25

Reverse folds.

26

Fold the inside layers together
for this spine-lock fold. Turn
over and repeat.

27

Fold the edge to the dot.
Turn over and repeat.

28

1. Lift up at the dot.
2. Fold under.
3. Bring the ends
   together.
Turn over and repeat.

29

Only the top half is drawn.
Tuck the tab under A. B
goes into the lower half
when the same folds are
repeated below.

30

Hexagonal Dipyramid
in a Sphere

Squat Silver Hexagonal Dipyramid

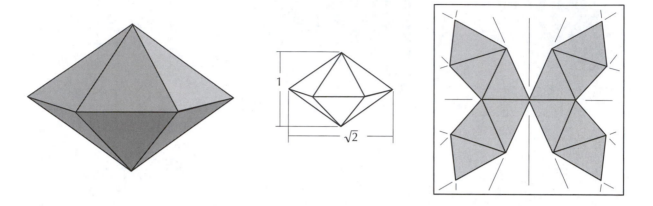

The ratio of the height of this hexagonal dipyramid to its diameter is 1 to $\sqrt{2}$. The small angle in each triangle is about 48°, same as in the Silver Square Dipyramid. The folding is similar to the Hexagonal Dipyramid in a Sphere.

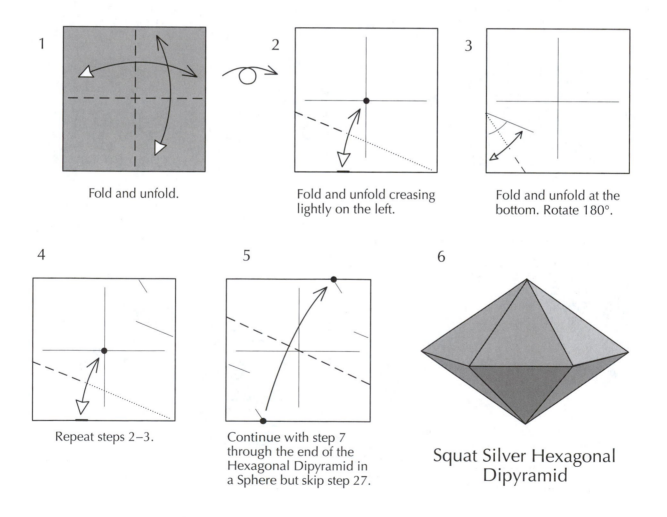

1 Fold and unfold.

2 Fold and unfold creasing lightly on the left.

3 Fold and unfold at the bottom. Rotate 180°.

4 Repeat steps 2–3.

5 Continue with step 7 through the end of the Hexagonal Dipyramid in a Sphere but skip step 27.

6 Squat Silver Hexagonal Dipyramid

Heptagonal Dipyramid

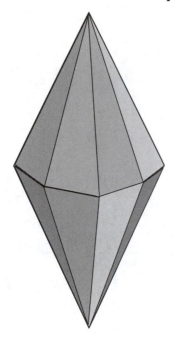

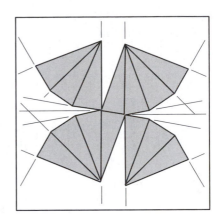

This fourteen-sided dipyramid is the dual of the uniform heptagonal prism. The small angle in each triangle is about 22°.

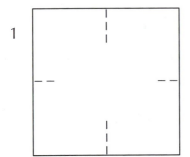

1

Fold and unfold in half on the edges. Make longer vertical creases.

2

Crease at the bottom.

3

Unfold.

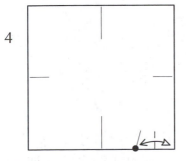

4

Fold and unfold at the bottom.

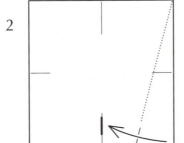

5

Fold and unfold. Rotate 180°.

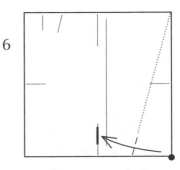

6

Repeat steps 2–5.

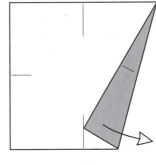

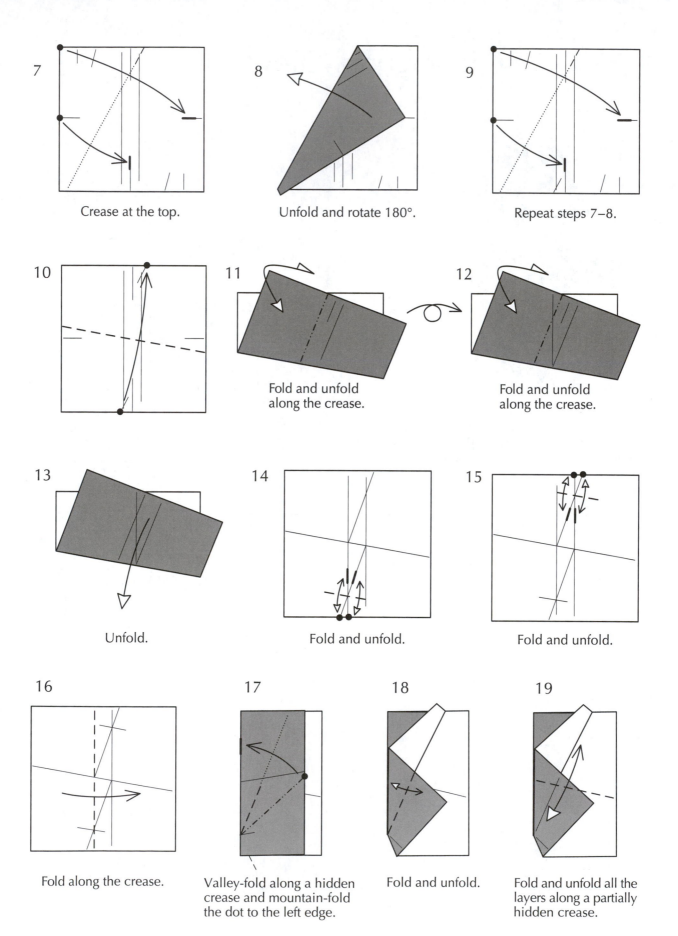

7

Crease at the top.

8

Unfold and rotate 180°.

9

Repeat steps 7–8.

10

11

Fold and unfold
along the crease.

12

Fold and unfold
along the crease.

13

Unfold.

14

Fold and unfold.

15

Fold and unfold.

16

Fold along the crease.

17

Valley-fold along a hidden
crease and mountain-fold
the dot to the left edge.

18

Fold and unfold.

19

Fold and unfold all the
layers along a partially
hidden crease.

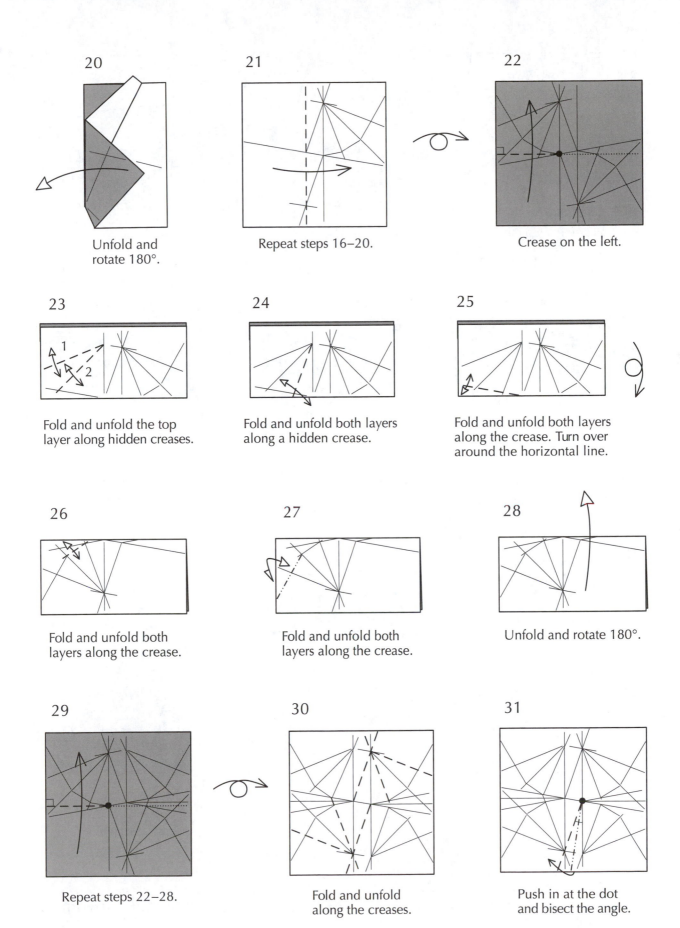

20

Unfold and
rotate 180°.

21

Repeat steps 16–20.

22

Crease on the left.

23

Fold and unfold the top
layer along hidden creases.

24

Fold and unfold both layers
along a hidden crease.

25

Fold and unfold both layers
along the crease. Turn over
around the horizontal line.

26

Fold and unfold both
layers along the crease.

27

Fold and unfold both
layers along the crease.

28

Unfold and rotate 180°.

29

Repeat steps 22–28.

30

Fold and unfold
along the creases.

31

Push in at the dot
and bisect the angle.

32

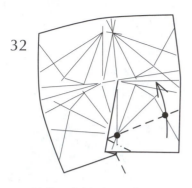

Valley-fold along the crease between the dots for this squash fold. Rotate 180°.

33

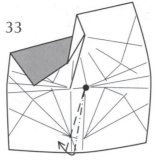

Repeat steps 31–32. Then flatten.

34

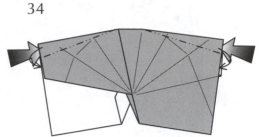

Reverse folds.

35

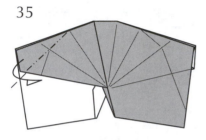

Fold the inside layers together for this spine-lock fold. Turn over and repeat.

36

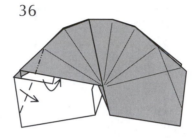

Fold the inside layers together for this spine-lock fold. Turn over and repeat.

37

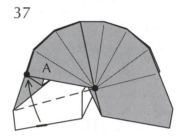

Bring the edge to the dot. All of the folds will be under region A. Turn over and repeat.

38

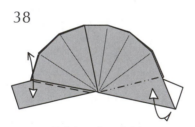

Fold and unfold.

39

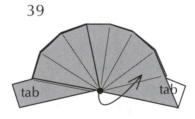

Lift up at the dot to open the model. Tuck and interlock the tabs.

40

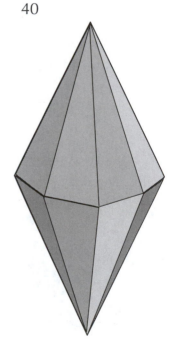

Heptagonal Dipyramid

Heptagonal Dipyramid 30°

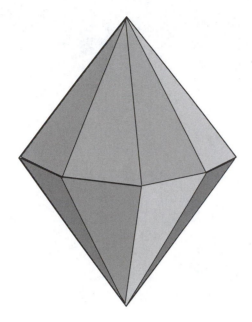

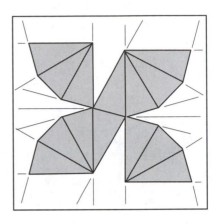

This fourteen-sided dipyramid is similar to the Heptagonal Dipyramid. The small angle in each triangle is 30°, which simplifies the geometry.

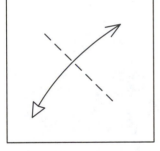

1

Fold and unfold creasing lightly.

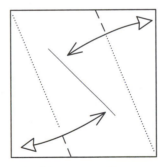

2

Fold and unfold at the top and bottom.

3

Fold and unfold.

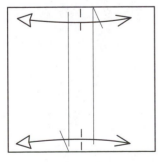

4

Fold and unfold in half at the top and bottom.

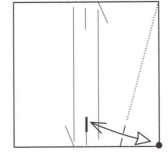

5

Fold and unfold at bottom.

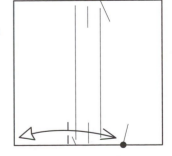

6

Fold and unfold at bottom. Rotate 180°.

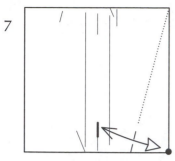

7

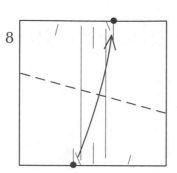

8

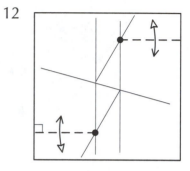

9

Repeat steps 5–6.

Fold and unfold
along the crease.

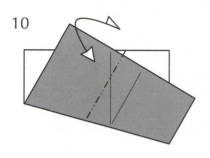

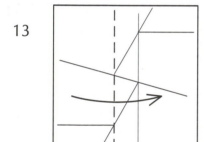

10

Fold and unfold
along the crease.

11

Unfold.

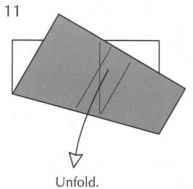

12

Fold and unfold.

13

Continue with step 16 through the end
of the Heptagonal Dipyramid. Skip step
18 since the crease is already there.

14

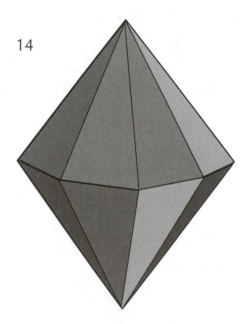

Heptagonal Dipyramid 30°

Heptagonal Dipyramid in a Sphere

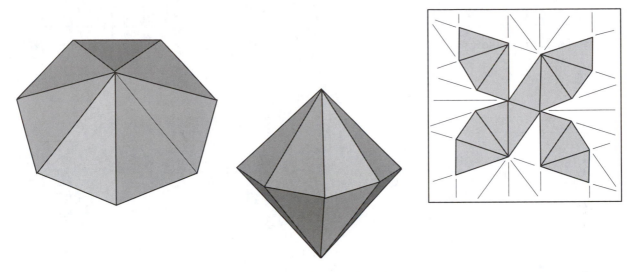

This dipyramid is inscibed in a sphere. The angles of each of the fourteen triangles are 36°, 72°, and 72°.

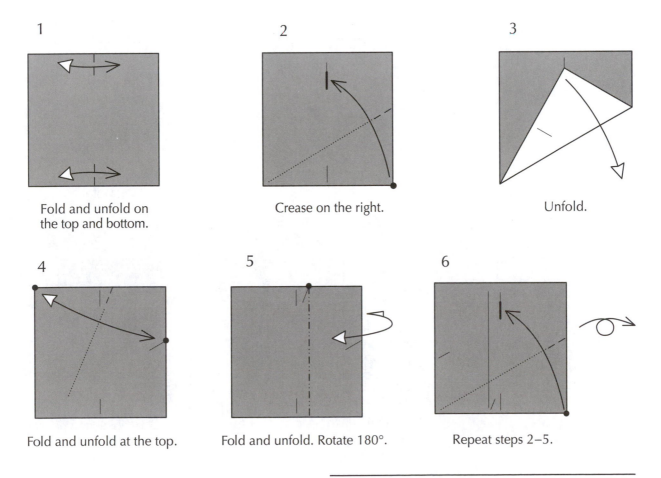

1

Fold and unfold on
the top and bottom.

2

Crease on the right.

3

Unfold.

4

Fold and unfold at the top.

5

Fold and unfold. Rotate 180°.

6

Repeat steps 2–5.

7

Fold and unfold
on the right.

8

Fold and unfold on
the top. Rotate 180°.

9

Repeat steps 7–8.

10

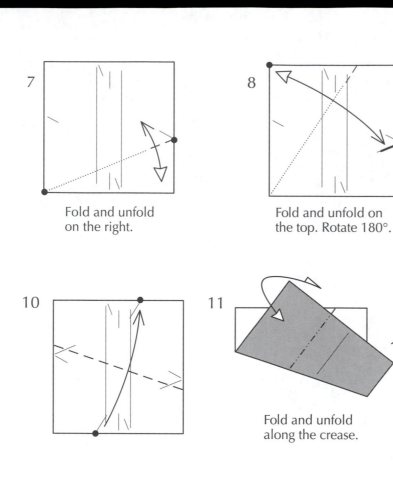

11

Fold and unfold
along the crease.

12

Fold and unfold
along the crease.

13

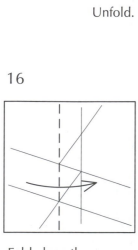

Unfold.

14

Fold and unfold.

15

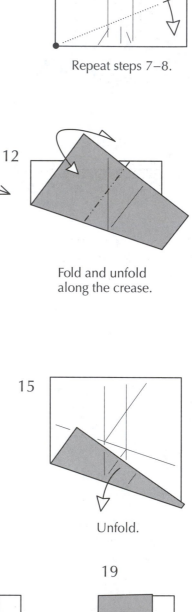

Unfold.

16

Fold along the crease.

17

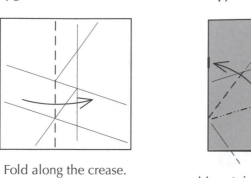

Mountain-fold
along the crease.

18

Fold and unfold.

19

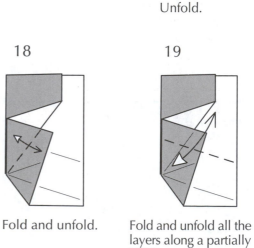

Fold and unfold all the
layers along a partially
hidden crease.

20

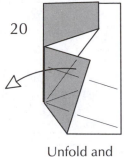

Unfold and
rotate 180°.

21

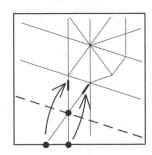

Repeat steps 14–20.

22

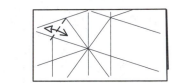

Crease on the left.

23

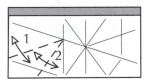

Fold and unfold the top
layer along hidden creases.

24

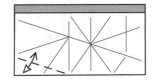

Fold and unfold both layers
along the crease. Turn over
around the horizontal line.

25

Fold and unfold both
layers along the crease.

26

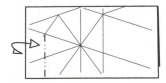

Fold and unfold both
layers along the crease.

27

Unfold and rotate 180°.

28

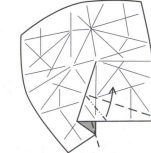

Repeat steps 22–27.

29

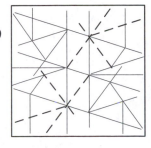

Fold and unfold
along the creases.

30

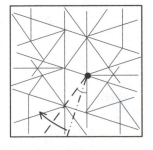

Push in at the dot
and bisect the angle.

31

Squash-fold.

32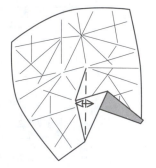

Fold and unfold.

33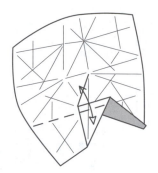

Fold and unfold.
Rotate 180°.

34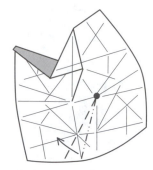

Repeat steps 30–33.
Then flatten.

35

Reverse folds.

36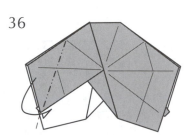

Fold the inside layers
together for this spine-lock
fold. Turn over and repeat.

37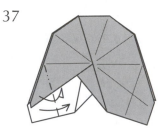

Fold the inside layers
together for this spine-lock
fold. Turn over and repeat.

38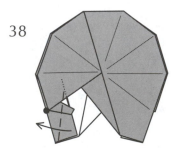

Fold the edge to the dot.
Turn over and repeat.

39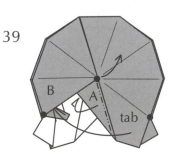

To finish the dipyramid:
1. Lift up at the upper dot.
2. Fold A inside.
3. Bring the tab behind B
   so the dots meet.
4. Repeat on the other side and
   tuck and interlock the tabs.

40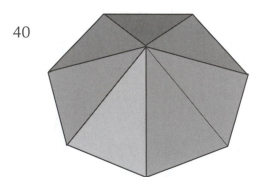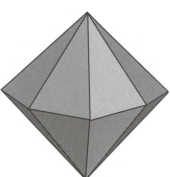

Heptagonal Dipyramid
in a Sphere

Octagonal Dipyramid

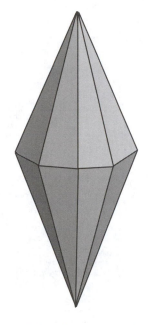

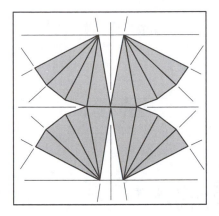

This 16-sided dipyramid is the dual of the uniform octagonal prism. The small angle in each triangle is about 17°. The ratio of the height to the diameter is $(1 + \sqrt{2})$ to 1.

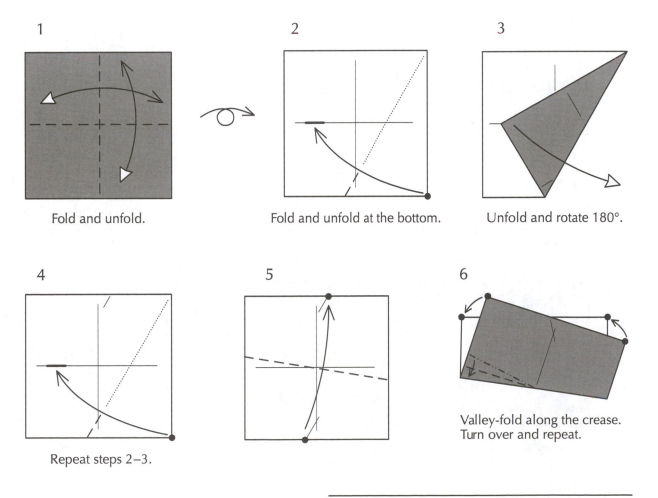

1

Fold and unfold.

2

Fold and unfold at the bottom.

3

Unfold and rotate 180°.

4

Repeat steps 2–3.

5

6

Valley-fold along the crease. Turn over and repeat.

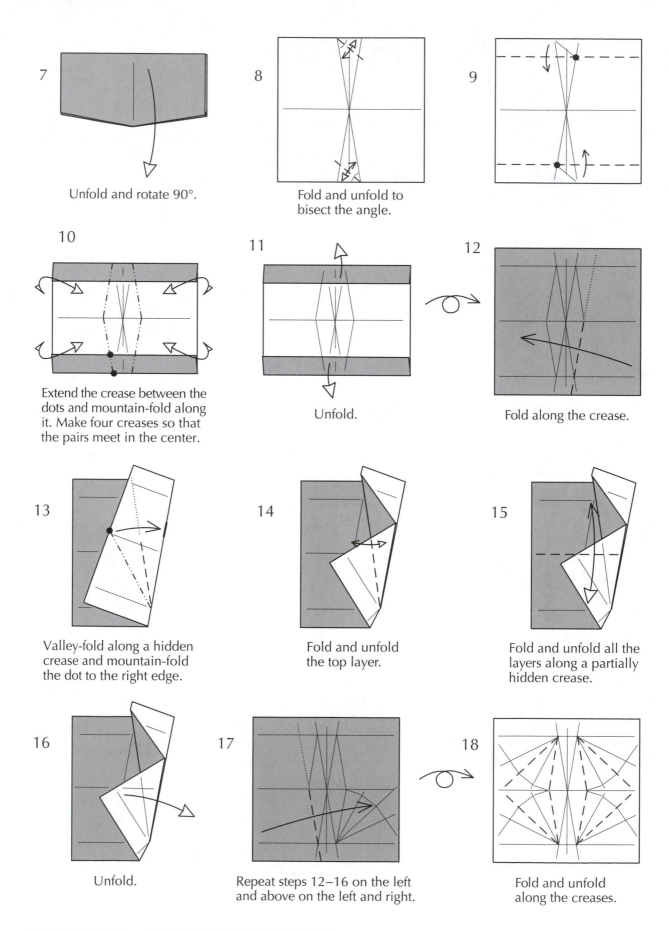

7

Unfold and rotate 90°.

8

Fold and unfold to bisect the angle.

9

10

Extend the crease between the dots and mountain-fold along it. Make four creases so that the pairs meet in the center.

11

Unfold.

12

Fold along the crease.

13

Valley-fold along a hidden crease and mountain-fold the dot to the right edge.

14

Fold and unfold the top layer.

15

Fold and unfold all the layers along a partially hidden crease.

16

Unfold.

17

Repeat steps 12–16 on the left and above on the left and right.

18

Fold and unfold along the creases.

19

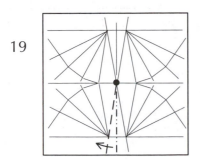

Push in at the dot.

20

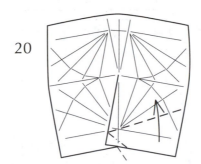

Squash-fold and rotate 180°.

21

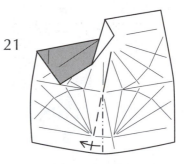

Repeat steps 19–20.
Then flatten.

22

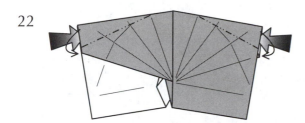

Reverse folds.

23

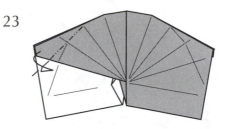

Fold the inside layers together
for this spine-lock fold. Turn
over and repeat.

24

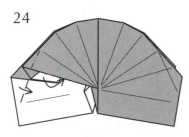

Fold the inside layers together
for this spine-lock fold. Turn
over and repeat.

25

Bring the edge to the dot. All of
the folds will be under region A.
Turn over and repeat.

26

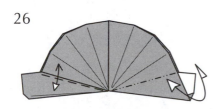

Fold and unfold.

27

Lift up at the dot to
open the model. Tuck
and interlock the tabs.

28

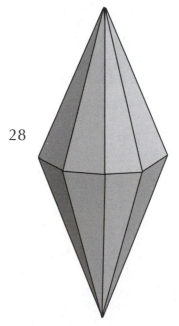

Octagonal Dipyramid

Octagonal Dipyramid 26°

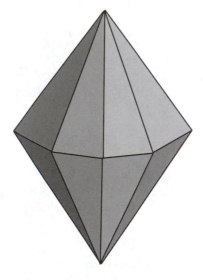

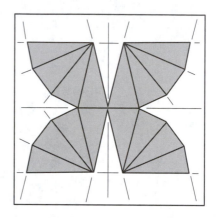

The folding of this 16-sided dipyramid is similar to the Octagonal Dipyramid. The small angle in each triangle is about 25.7°. That angle makes the geometry convenient since the horizontal lines are part of the crease pattern.

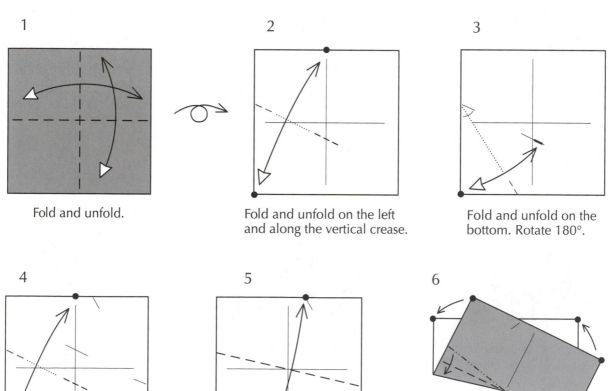

1

Fold and unfold.

2

Fold and unfold on the left and along the vertical crease.

3

Fold and unfold on the bottom. Rotate 180°.

4

Repeat steps 2–3.

5

6

Valley-fold along the crease. Turn over and repeat.

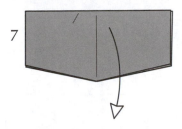

7

Unfold and rotate 90°.

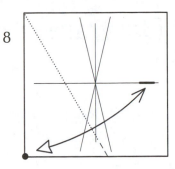

8

Fold and unfold at the bottom.

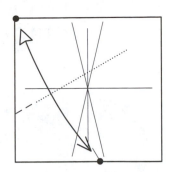

9

Fold and unfold on the left.

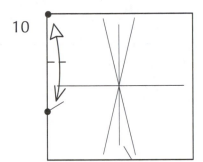

10

Fold and unfold on the left.

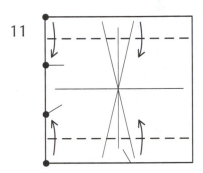

11

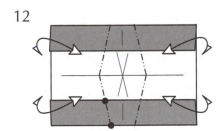

12

Continue with step 10 through the end of the Octagonal Dipyramid. Skip step 14 since the crease is already there. In step 20, fold along the lowest crease on the right.

13

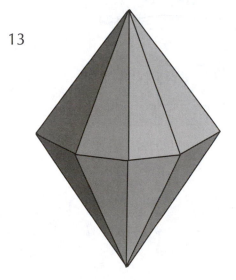

Octagonal Dipyramid 26°

Octagonal Dipyramid in a Sphere

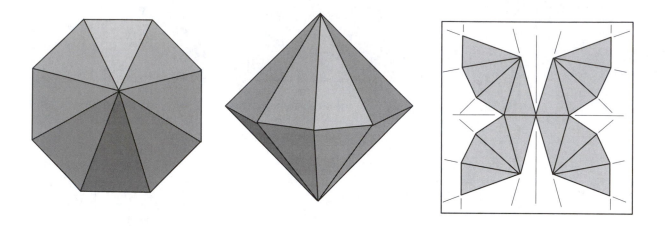

This dipyramid is inscibed in a sphere. The angles of each of the 16 triangles are 31.4°, 74.3°, and 74.3°.

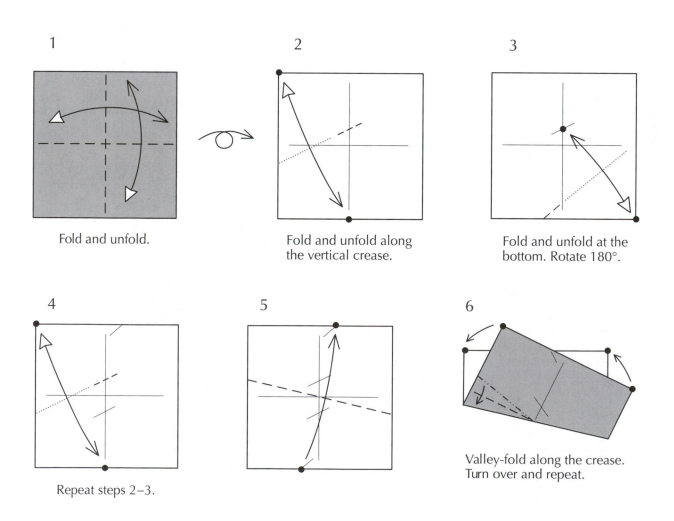

1

Fold and unfold.

2

Fold and unfold along the vertical crease.

3

Fold and unfold at the bottom. Rotate 180°.

4

Repeat steps 2–3.

5

6

Valley-fold along the crease. Turn over and repeat.

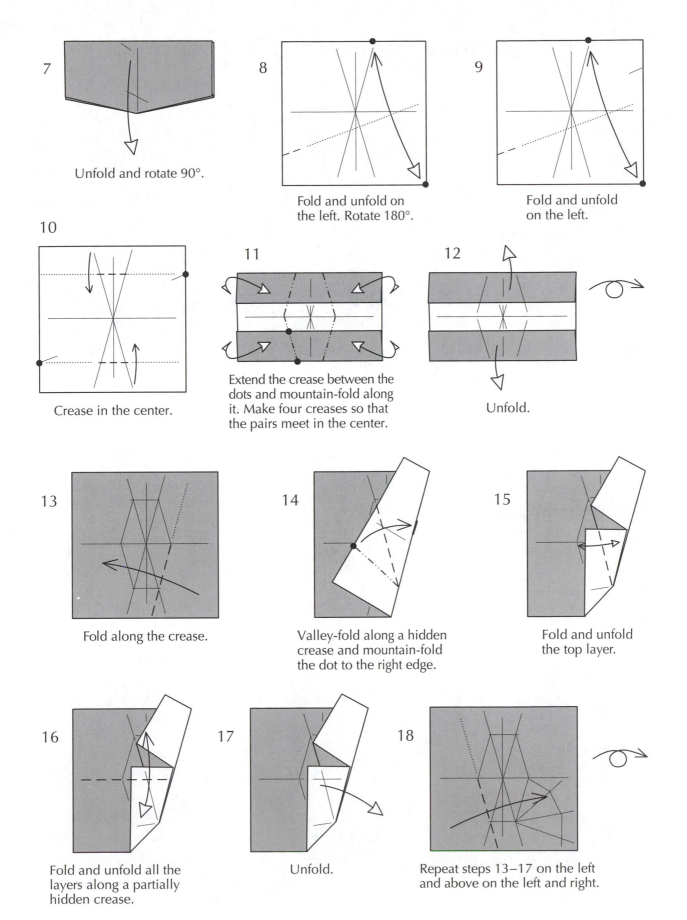

7

Unfold and rotate 90°.

8

Fold and unfold on
the left. Rotate 180°.

9

Fold and unfold
on the left.

10

Crease in the center.

11

Extend the crease between the
dots and mountain-fold along
it. Make four creases so that
the pairs meet in the center.

12

Unfold.

13

Fold along the crease.

14

Valley-fold along a hidden
crease and mountain-fold
the dot to the right edge.

15

Fold and unfold
the top layer.

16

Fold and unfold all the
layers along a partially
hidden crease.

17

Unfold.

18

Repeat steps 13–17 on the left
and above on the left and right.

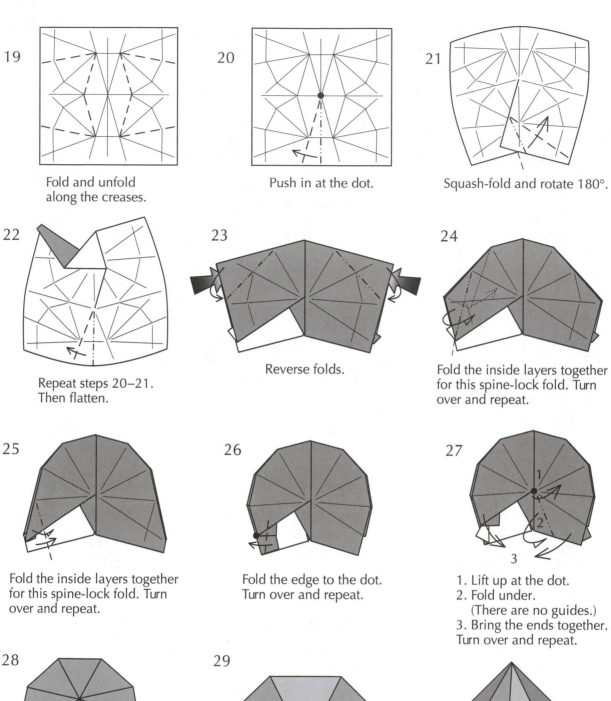

19 Fold and unfold along the creases.

20 Push in at the dot.

21 Squash-fold and rotate 180°.

22 Repeat steps 20–21. Then flatten.

23 Reverse folds.

24 Fold the inside layers together for this spine-lock fold. Turn over and repeat.

25 Fold the inside layers together for this spine-lock fold. Turn over and repeat.

26 Fold the edge to the dot. Turn over and repeat.

27
1. Lift up at the dot.
2. Fold under. (There are no guides.)
3. Bring the ends together. Turn over and repeat.

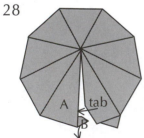

28 Only the top half is drawn. Tuck the tab under A. B goes into the lower half while the same folds are repeated below.

29

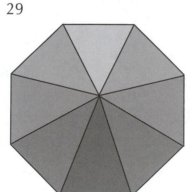

Octagonal Dipyramid in a Sphere

Nonagonal Dipyramid

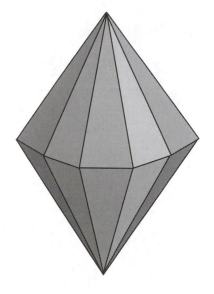

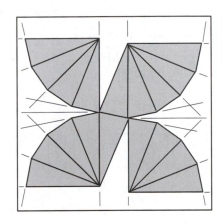

The angle at the top of the triangle in this 18-sided dipyramid is 22.5°.

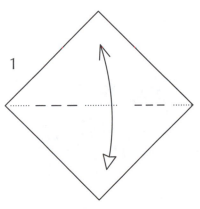

1

Fold and unfold
creasing lightly. Rotate.

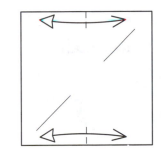

2

Fold and unfold on
the top and bottom.

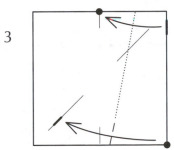

3

Bring the right edge to the upper
dot and the lower dot to the
diagonal. Crease on the bottom.

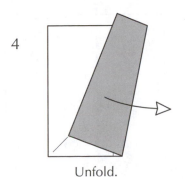

4

Unfold.

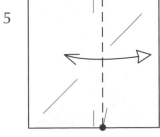

5

Fold and unfold.
Rotate 180°.

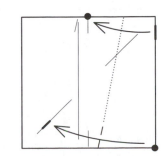

6

Repeat steps 3–5.

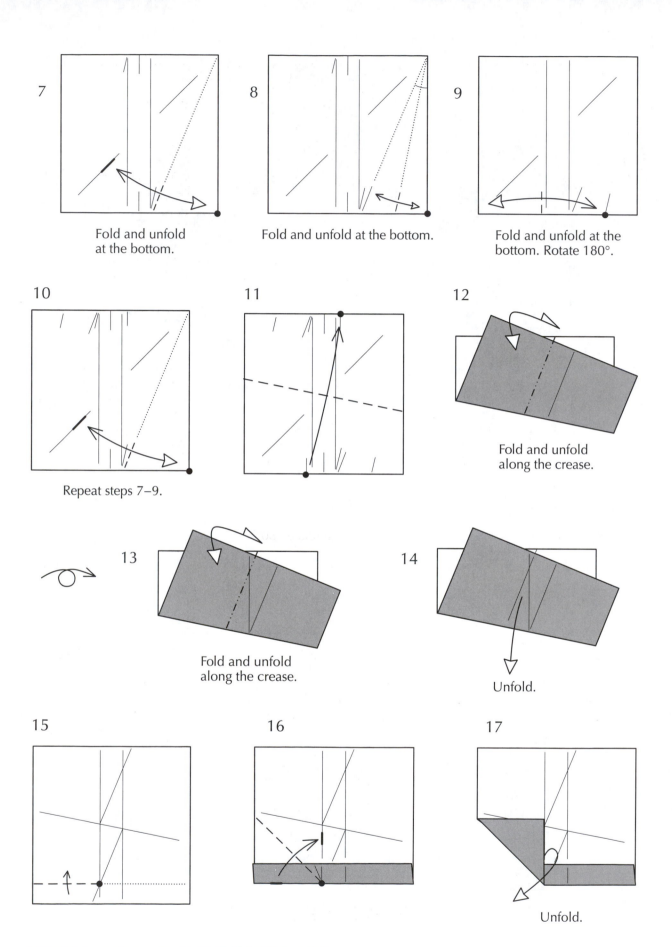

7

Fold and unfold
at the bottom.

8

Fold and unfold at the bottom.

9

Fold and unfold at the
bottom. Rotate 180°.

10

Repeat steps 7–9.

11

12

Fold and unfold
along the crease.

13

Fold and unfold
along the crease.

14

Unfold.

15

16

17

Unfold.

18

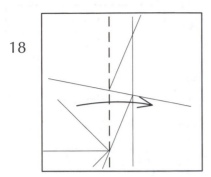

Fold along the crease.

19

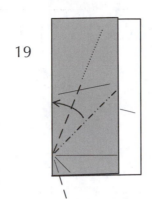

Mountain-fold along the crease and bring it to the left edge.

20

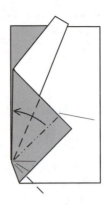

Mountain-fold along the crease and bring it to the left edge.

21

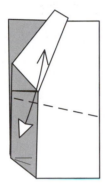

Fold and unfold all the layers along a partially hidden crease.

22

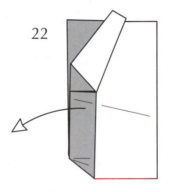

Unfold and rotate 180°. Repeat steps 15–21.

Now you are on your own. The folding is similar to that of the Heptagonal Dipyramid.

23

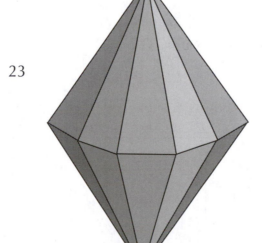

Nonagonal Dipyramid

Decagonal Dipyramid

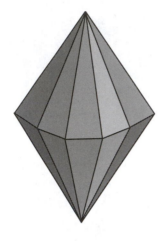
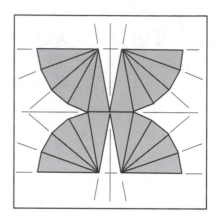

The angle at the top of the triangles in this 20-sided dipyramid is 20°.

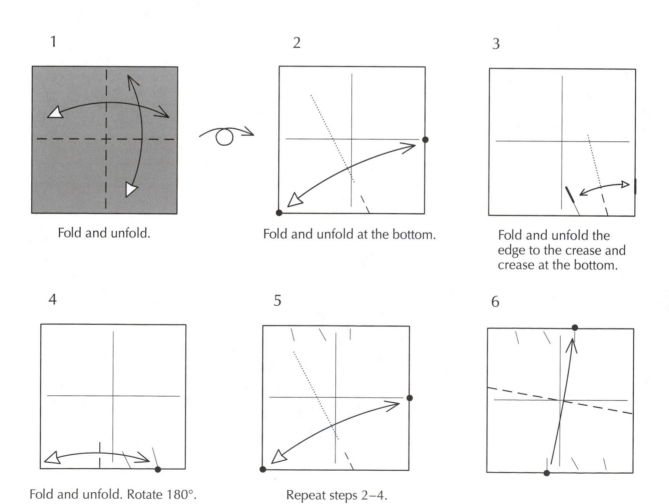

1

Fold and unfold.

2

Fold and unfold at the bottom.

3

Fold and unfold the edge to the crease and crease at the bottom.

4

Fold and unfold. Rotate 180°.

5

Repeat steps 2–4.

6

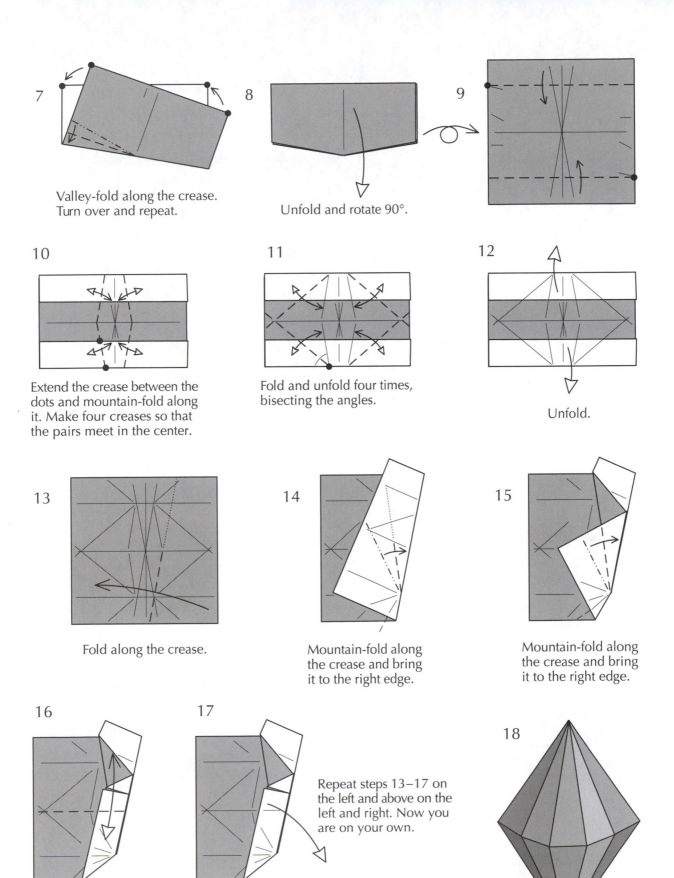

7 Valley-fold along the crease. Turn over and repeat.

8 Unfold and rotate 90°.

9

10 Extend the crease between the dots and mountain-fold along it. Make four creases so that the pairs meet in the center.

11 Fold and unfold four times, bisecting the angles.

12 Unfold.

13 Fold along the crease.

14 Mountain-fold along the crease and bring it to the right edge.

15 Mountain-fold along the crease and bring it to the right edge.

16 Fold and unfold all the layers.

17 Unfold.

Repeat steps 13–17 on the left and above on the left and right. Now you are on your own.

18

Decagonal Dipyramid

Dimpled Dipyramid Design

Dimpled dipyramids are polyhedra where alternate sides of the dipyramids are sunken. They are each composed of three different triangular faces. Triangle A represents the flat faces (isosceles triangles), which are the same as the faces of the nondimpled models. Triangles B are the (typically smaller) isosceles triangles by the equator. Triangles C are the longer triangles that complete the sunken faces. Formulas from "Dipyramid Design" (page 173) will be used to relate the proportion of the height to the diameter to the angle α of triangle A.

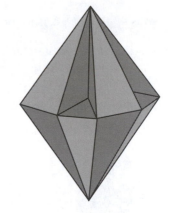

Given polygon n and angle α, the proportion of the height of the dipyramid to its diameter, H, is

$$H = \tan\left(\arccos\left(\frac{\sin(\alpha/2)}{\sin(180°/n)}\right)\right)$$

Given H and polygon n, α is found by

$$\alpha = 2\arcsin[\sin(180°/n)\cos(\arctan(H))]$$

For a dimpled dipyramid with a given polygon base n, and angle α in triangle A, the triangles B and C can vary. The models in this section are made with square, hexagonal, and octagonal bases. (Note that, in general, n must be even.) Since these shapes are complex and thus difficult to design, I worked out methods to define triangles B and C, given the polygon base, so as to simplify the folding.

    Here is a trio of dimpled dipyramids where angle α is the same for all (α = 36.87°).

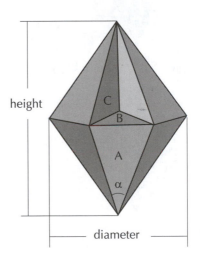

height

diameter

Tall Dimpled Square Dipyramid
$H = 2$

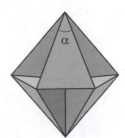

Dimpled Hexagonal Dipyramid
$H = \sqrt{1.5}$

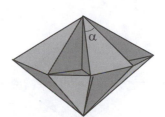

Octagonal Flying Saucer
$H = 1/\sqrt{2}$

Square Dimpled Dipyramids

For these models, triangle B is chosen to be a 45° right triangle. The crease patterns have odd symmetry.

Since *n* is 4, given *H*, α is found by

$$\alpha = 2\arcsin[\cos(\arctan(H))/\sqrt{2}\,]$$

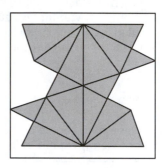

Tall Dimpled Square Dipyramid, *H* = 2, α = 36.87°

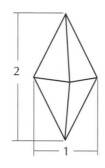

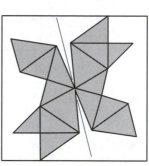

Dimpled Silver Square Dipyramid, H = √2̄, α = 48.19°

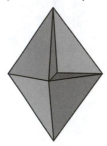

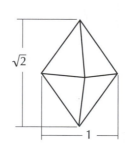

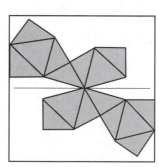

Heptahedron, H = 1, α = 60°

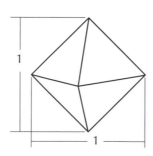

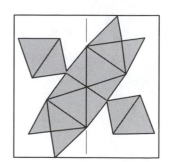

Dimpled Squat Square Dipyramid, H = 1/√2̄, a = 70.53°

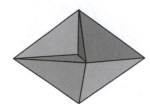

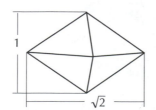

Hexagonal Dimpled Dipyramids

To simplify the folding for these models, triangles B and C are chosen by the following:

1. Draw triangle A twice as shown in the first picture.

2. Draw a line connecting the dots. Given angle α, angles γ and δ can be found.

3. Draw a line at angle δ from the base of triangle A. Thus triangles B and C are now determined.

The crease patterns have odd symmetry. Since n is 6, given H, α is found by

$$\alpha = 2\arcsin[\cos(\arctan(H))/2]$$

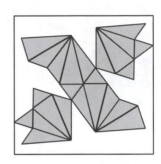

Tall Dimpled Hexagonal Dipyramid, $H = \sqrt{3}$, $\alpha = 28.96°$

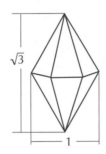
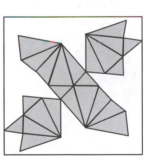

Dimpled Silver Hexagonal Dipyramid, $H = \sqrt{2}$, $\alpha = 33.56°$

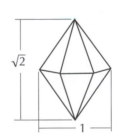
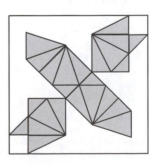

Dimpled Hexagonal Dipyramid, $H = \sqrt{1.5}$, $\alpha = 36.87°$

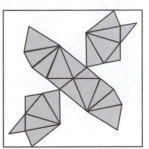

Dimpled Hexagonal Dipyramid in a Sphere, $H = 1$, $\alpha = 41.41°$

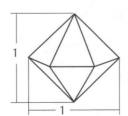

Octagonal Dimpled Dipyramids

To simplify the folding for these models, triangles B and C are chosen by the following:

1. Draw triangle A three times as shown in the first picture.

2. Add lines for angles γ such that α + 2γ = 90°.

3. Connect the three dots. Thus triangles B and C are now determined.

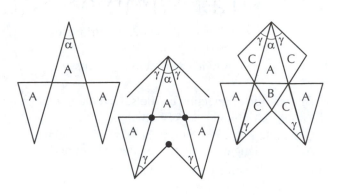

The crease patterns have square symmetry. Since n is 8, given H, α is found by

$$\alpha = 2\arcsin[\sin(22.5°)\cos(\arctan(H))]$$

Octagonal Flying Saucer, $H = 1/\sqrt{2}$, $\alpha = 36.87°$

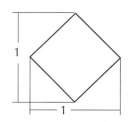
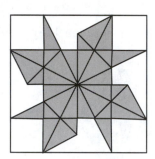

Dimpled Octagonal Dipyramid in a Sphere, $H = 1$, $\alpha = 31.4°$

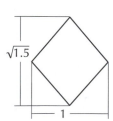
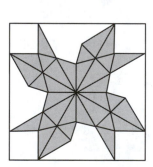

Dimpled Octagonal Dipyramid, $H = \sqrt{1.5}$, $\alpha = 28.96°$

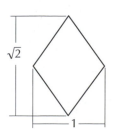
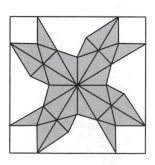

Dimpled Silver Octagonal Dipyramid, $H = \sqrt{2}$, $\alpha = 25.53°$

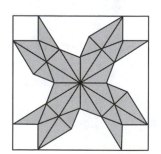

Tall Dimpled Square Dipyramid

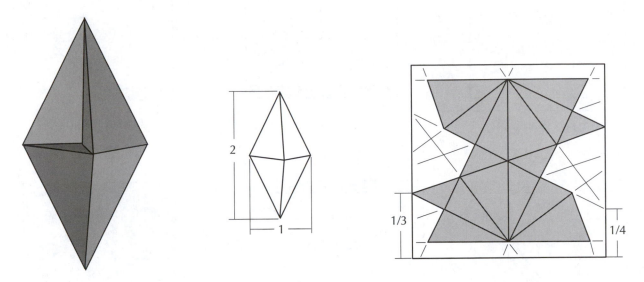

The height of this tall dimpled square dipyramid is twice the diameter. Four alternating sides are sunken. According to the formula I derived, the small angle in each nonsunken face is about 36.87° to achieve the dimensions.

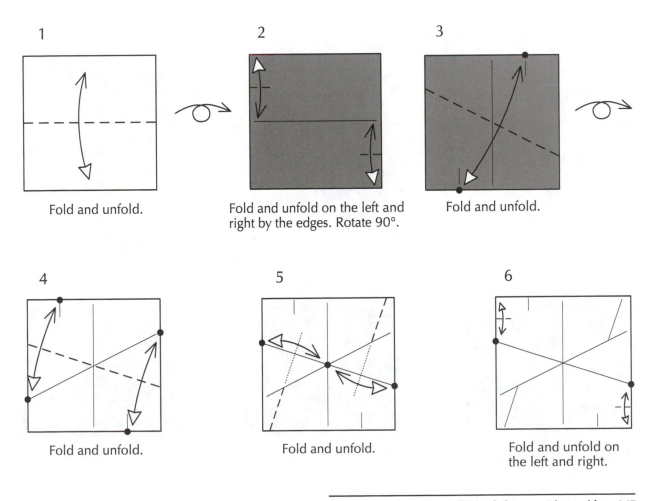

1

Fold and unfold.

2

Fold and unfold on the left and right by the edges. Rotate 90°.

3

Fold and unfold.

4

Fold and unfold.

5

Fold and unfold.

6

Fold and unfold on the left and right.

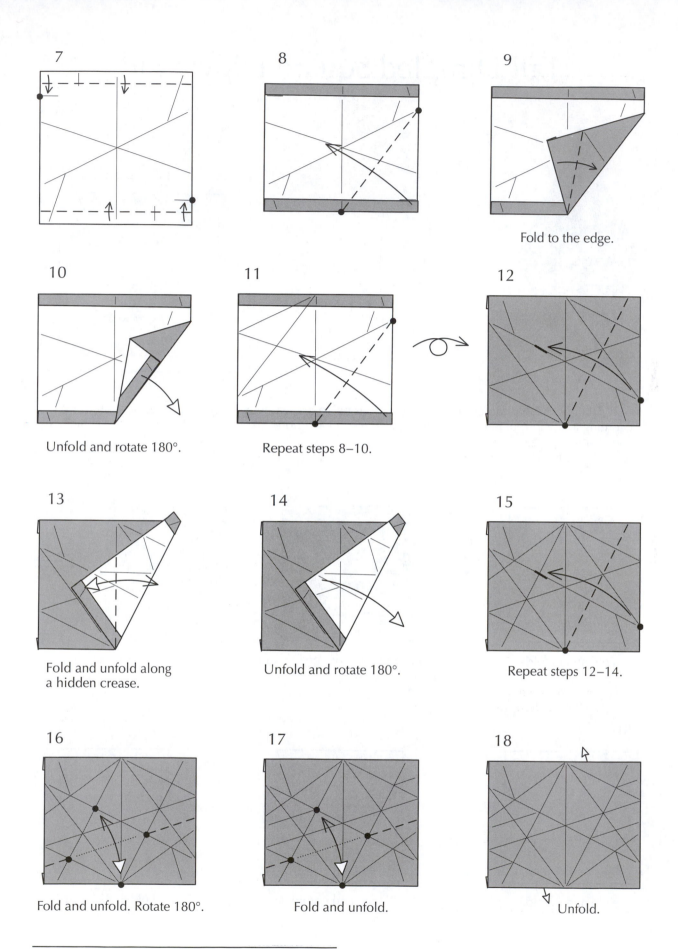

7

8

9

Fold to the edge.

10

Unfold and rotate 180°.

11

Repeat steps 8–10.

12

13

Fold and unfold along
a hidden crease.

14

Unfold and rotate 180°.

15

Repeat steps 12–14.

16

Fold and unfold. Rotate 180°.

17

Fold and unfold.

18

Unfold.

19

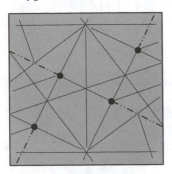

Fold and unfold along the creases. Rotate 90°.

20

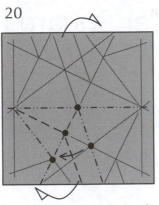

Puff out at the upper dot and push in at the center dot. The other two dots will meet. Flatten inside.

21

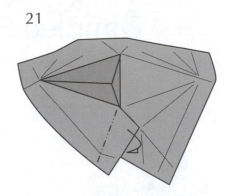

Tuck inside.

22

23

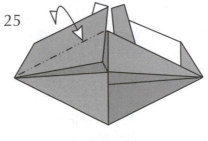

Rotate 180° and repeat steps 20–22. Rotate the bottom to the top.

24

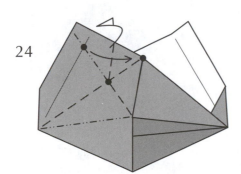

Push in at the lower dot. The other dots will meet. Turn over and repeat.

25

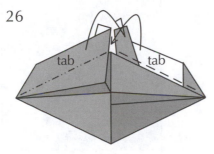

Fold and unfold. Turn over and repeat.

26

Interlock the tabs to close the model.

tab   tab

27

Tall Dimpled
Square Dipyramid

Dimpled Silver Square Dipyramid

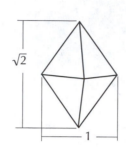

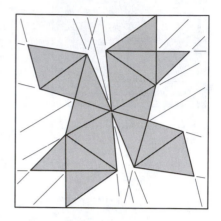

The ratio of the height of this square dipyramid to its width is $\sqrt{2}$ to 1. Four alternating sides are sunken. The small angle in each nonsunken face is about 48° to achieve the dimensions.

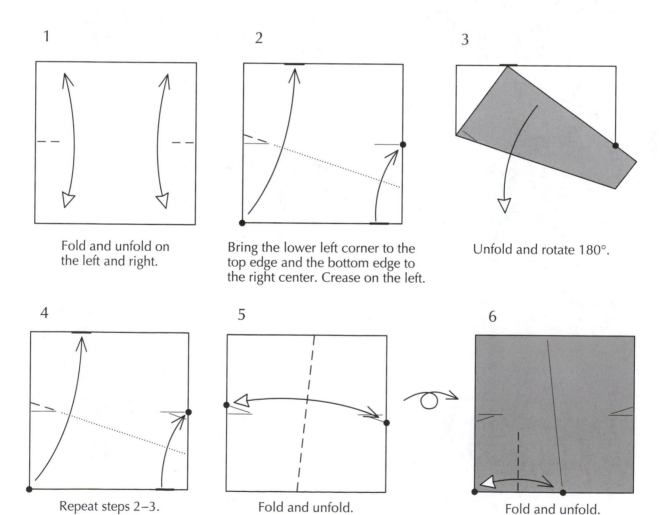

1

Fold and unfold on the left and right.

2

Bring the lower left corner to the top edge and the bottom edge to the right center. Crease on the left.

3

Unfold and rotate 180°.

4

Repeat steps 2–3.

5

Fold and unfold.

6

Fold and unfold.

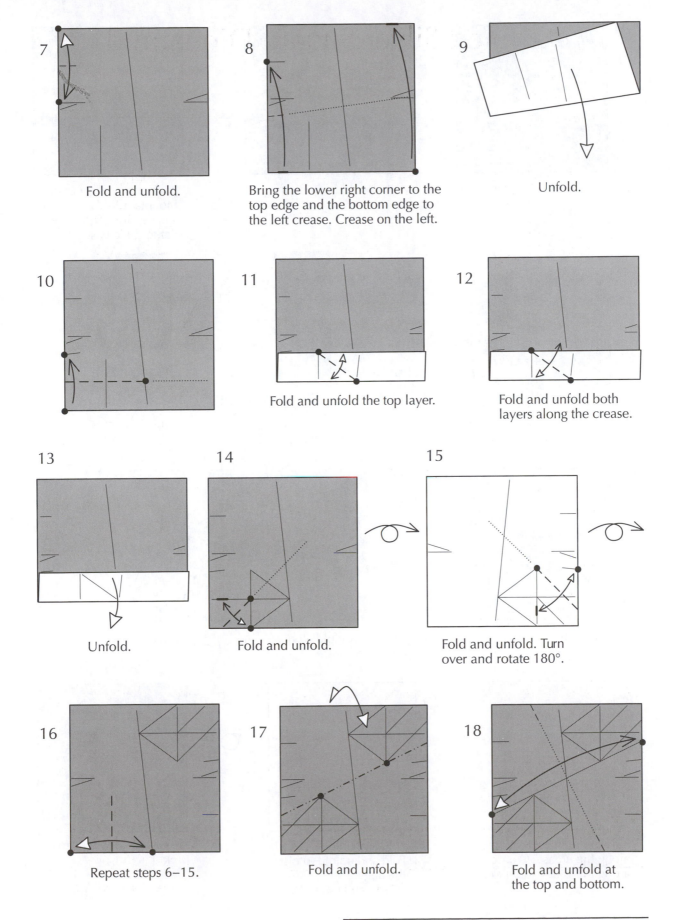

7 Fold and unfold.

8 Bring the lower right corner to the top edge and the bottom edge to the left crease. Crease on the left.

9 Unfold.

10

11 Fold and unfold the top layer.

12 Fold and unfold both layers along the crease.

13 Unfold.

14 Fold and unfold.

15 Fold and unfold. Turn over and rotate 180°.

16 Repeat steps 6–15.

17 Fold and unfold.

18 Fold and unfold at the top and bottom.

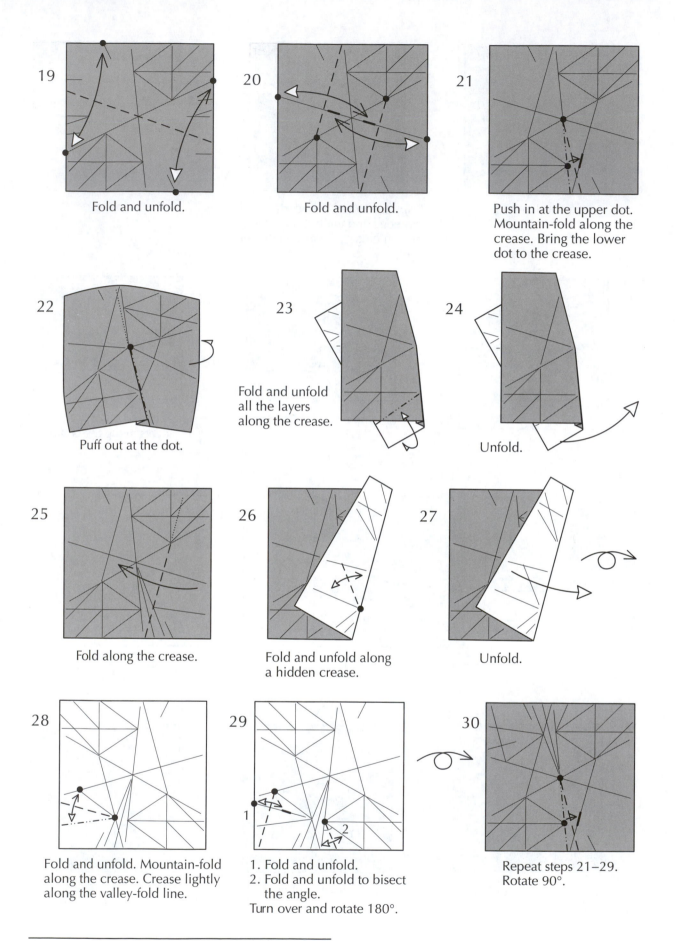

19

Fold and unfold.

20

Fold and unfold.

21

Push in at the upper dot. Mountain-fold along the crease. Bring the lower dot to the crease.

22

Puff out at the dot.

23

Fold and unfold all the layers along the crease.

24

Unfold.

25

Fold along the crease.

26

Fold and unfold along a hidden crease.

27

Unfold.

28

Fold and unfold. Mountain-fold along the crease. Crease lightly along the valley-fold line.

29

1. Fold and unfold.
2. Fold and unfold to bisect the angle.
Turn over and rotate 180°.

30

Repeat steps 21–29. Rotate 90°.

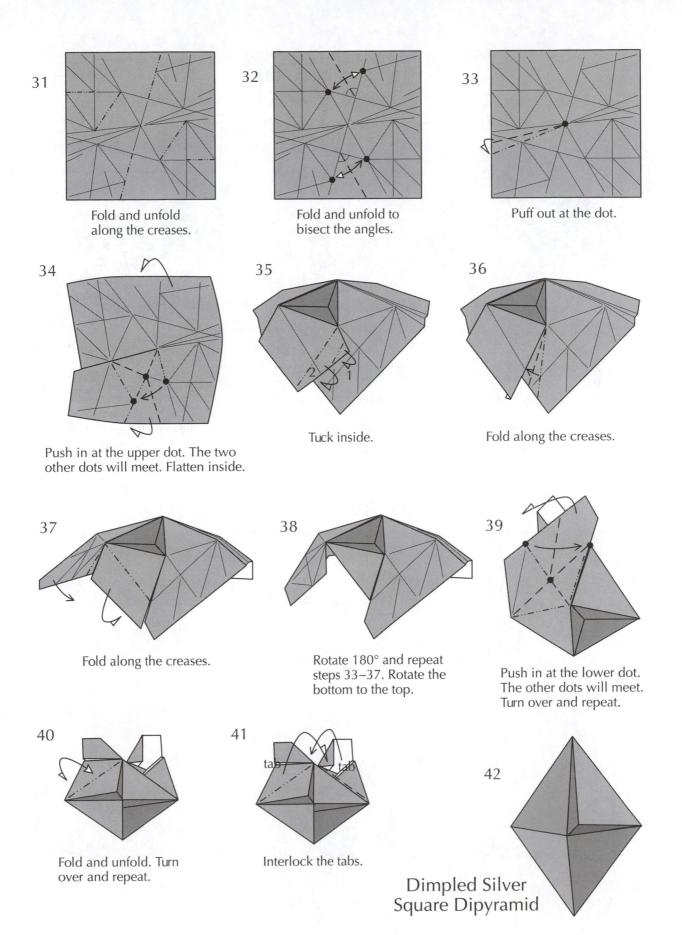

31 Fold and unfold along the creases.

32 Fold and unfold to bisect the angles.

33 Puff out at the dot.

34 Push in at the upper dot. The two other dots will meet. Flatten inside.

35 Tuck inside.

36 Fold along the creases.

37 Fold along the creases.

38 Rotate 180° and repeat steps 33–37. Rotate the bottom to the top.

39 Push in at the lower dot. The other dots will meet. Turn over and repeat.

40 Fold and unfold. Turn over and repeat.

41 Interlock the tabs.

42

Dimpled Silver Square Dipyramid

Heptahedron

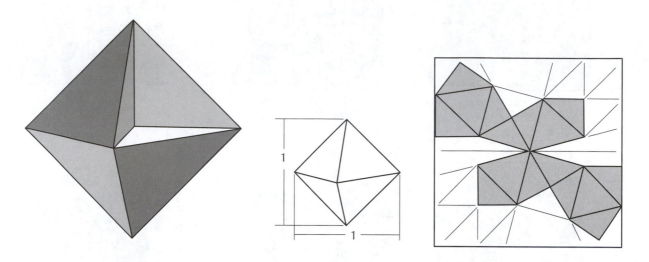

For this polyhedron, four faces are indented toward the center. It is named a heptahedron for its seven sides: four outer sides (same as the octahedron) and three center sides representing the x, y, and z axes. It can also be called a tetrahemihexahedron. This interesting shape combines equilateral triangles with isosceles right triangles.

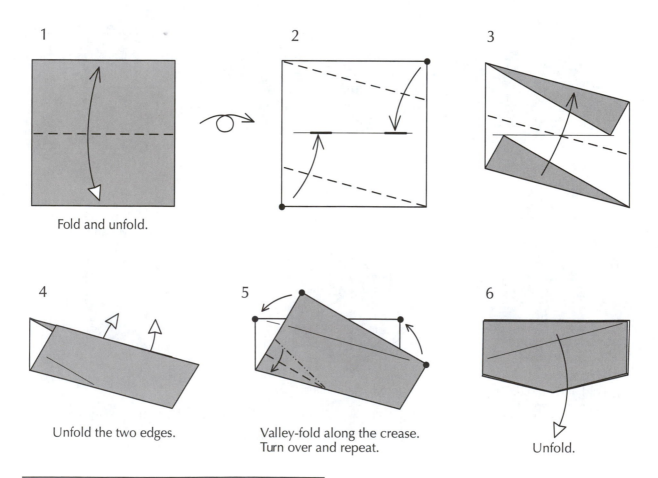

1

Fold and unfold.

2

3

4

Unfold the two edges.

5

Valley-fold along the crease.
Turn over and repeat.

6

Unfold.

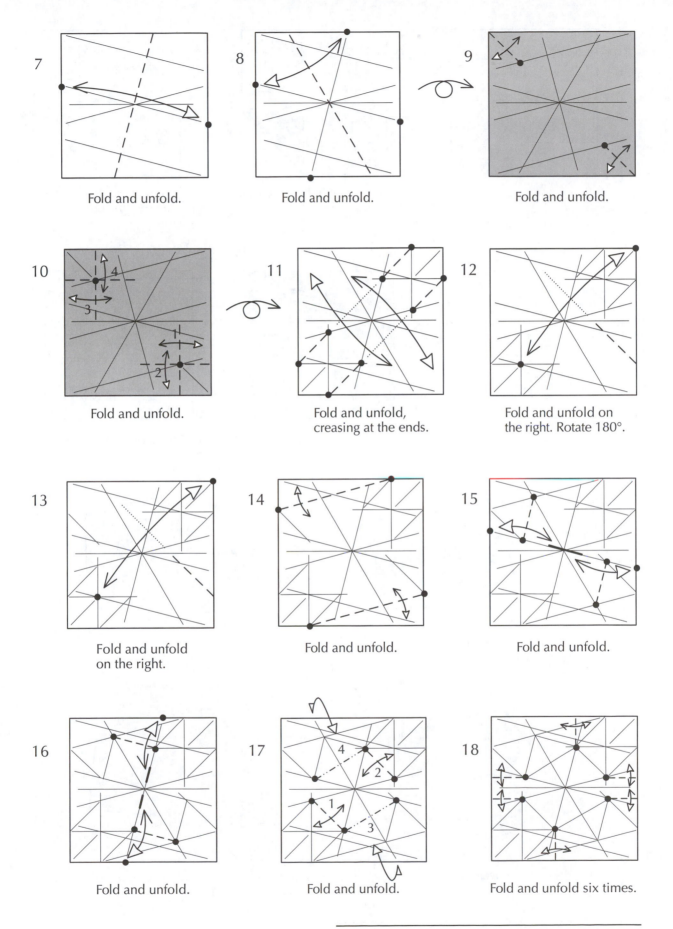

7 Fold and unfold.

8 Fold and unfold.

9 Fold and unfold.

10 Fold and unfold.

11 Fold and unfold,
creasing at the ends.

12 Fold and unfold on
the right. Rotate 180°.

13 Fold and unfold
on the right.

14 Fold and unfold.

15 Fold and unfold.

16 Fold and unfold.

17 Fold and unfold.

18 Fold and unfold six times.

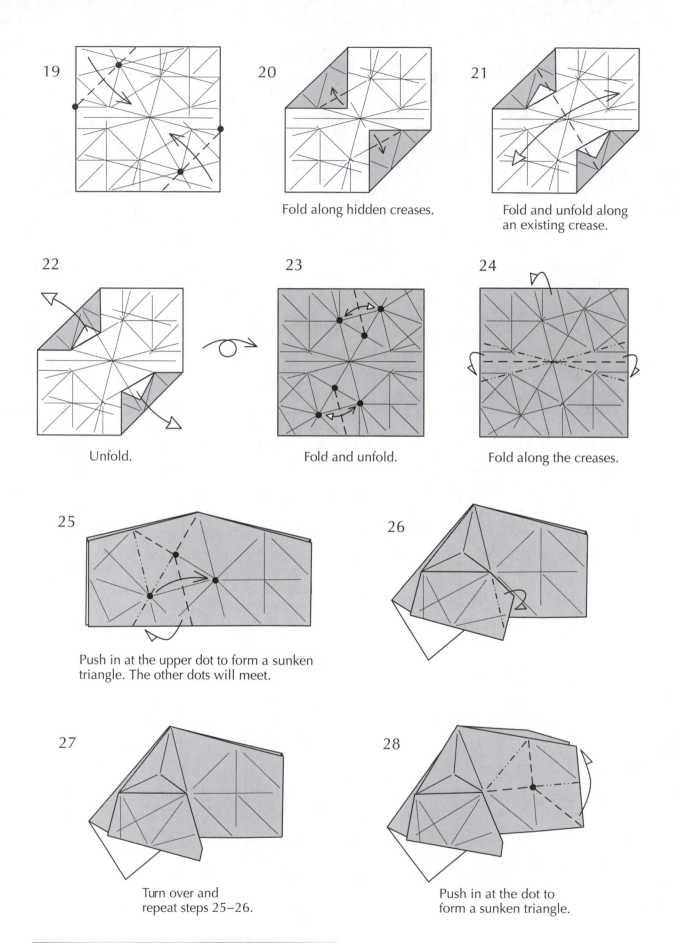

19

20

Fold along hidden creases.

21

Fold and unfold along
an existing crease.

22

Unfold.

23

Fold and unfold.

24

Fold along the creases.

25

Push in at the upper dot to form a sunken
triangle. The other dots will meet.

26

27

Turn over and
repeat steps 25–26.

28

Push in at the dot to
form a sunken triangle.

29

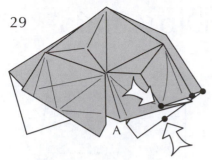

Note how some paper comes out at A. Flatten at the dots so each pair meets.

30

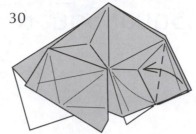

Fold all the layers together.

31

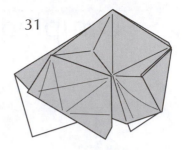

Turn over and repeat steps 28–30.

32

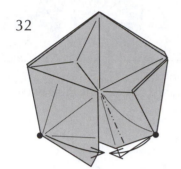

Bring the dots together. Rotate.

33

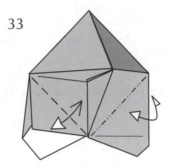

Flatten the flaps together. Fold and unfold all the layers.

34

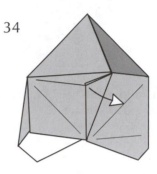

Unfold.

35

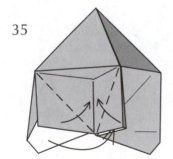

Tuck the left flap into the right ones and flatten.

36

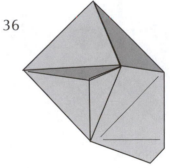

Turn over and repeat step 35.

37

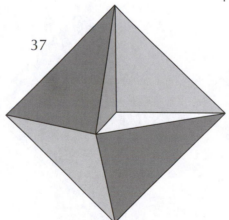

Heptahedron

Dimpled Squat Square Dipyramid

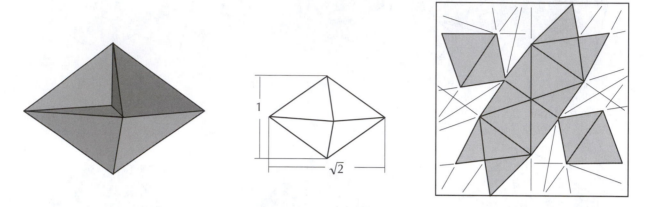

The ratio of the height of this dimpled squat square dipyramid to its diameter is 1 to $\sqrt{2}$. The crease pattern shows odd symmetry.

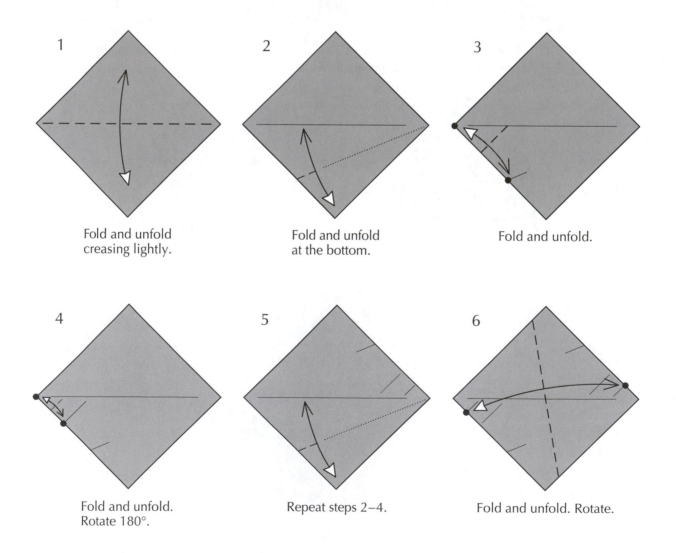

1

Fold and unfold
creasing lightly.

2

Fold and unfold
at the bottom.

3

Fold and unfold.

4

Fold and unfold.
Rotate 180°.

5

Repeat steps 2–4.

6

Fold and unfold. Rotate.

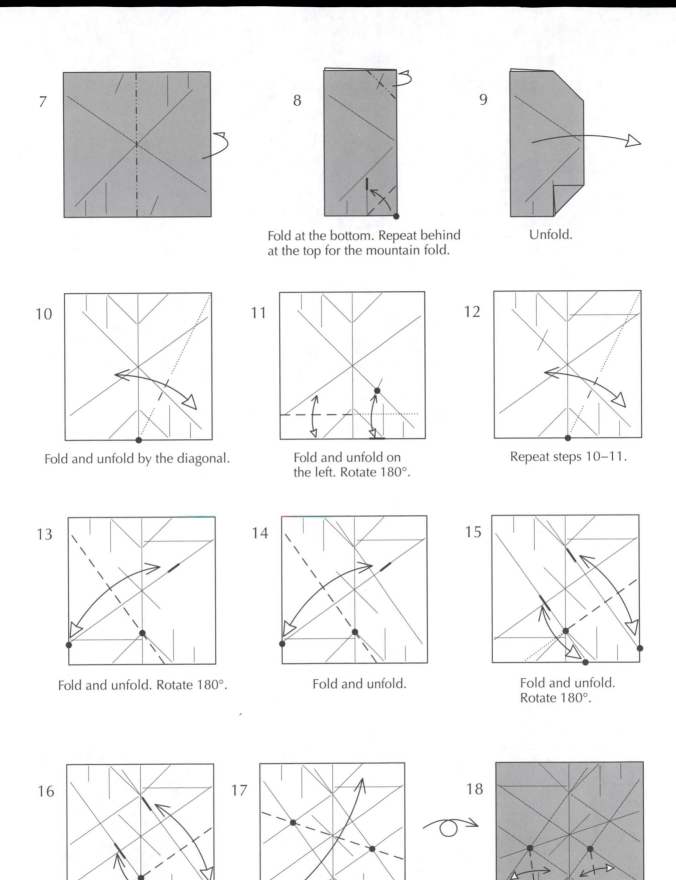

7

8

Fold at the bottom. Repeat behind at the top for the mountain fold.

9

Unfold.

10

Fold and unfold by the diagonal.

11

Fold and unfold on the left. Rotate 180°.

12

Repeat steps 10–11.

13

Fold and unfold. Rotate 180°.

14

Fold and unfold.

15

Fold and unfold. Rotate 180°.

16

Fold and unfold.

17

Fold and unfold.

18

Fold and unfold.

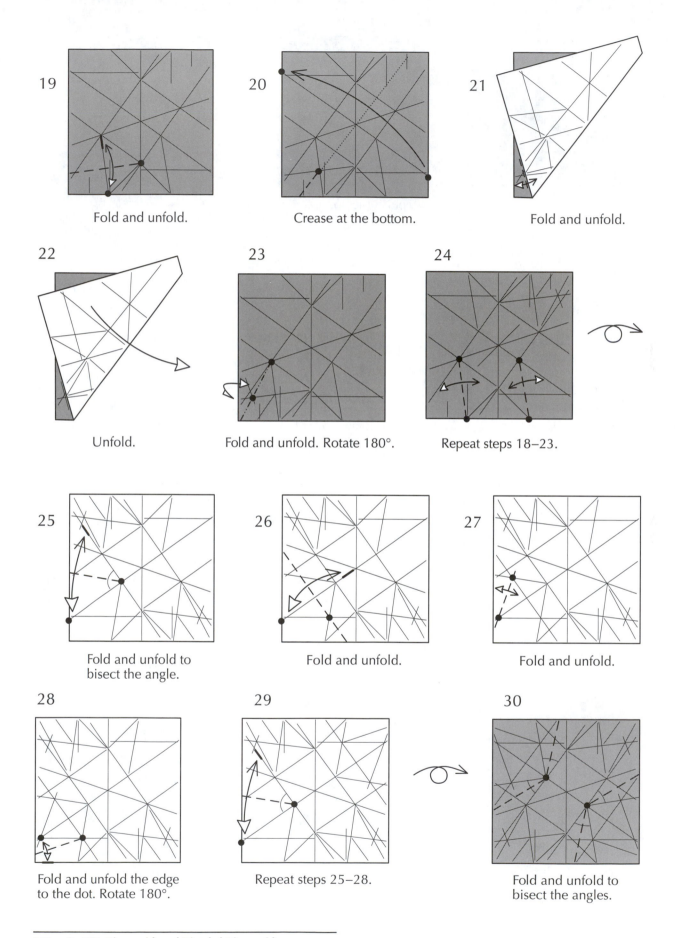

19

Fold and unfold.

20

Crease at the bottom.

21

Fold and unfold.

22

Unfold.

23

Fold and unfold. Rotate 180°.

24

Repeat steps 18–23.

25

Fold and unfold to bisect the angle.

26

Fold and unfold.

27

Fold and unfold.

28

Fold and unfold the edge to the dot. Rotate 180°.

29

Repeat steps 25–28.

30

Fold and unfold to bisect the angles.

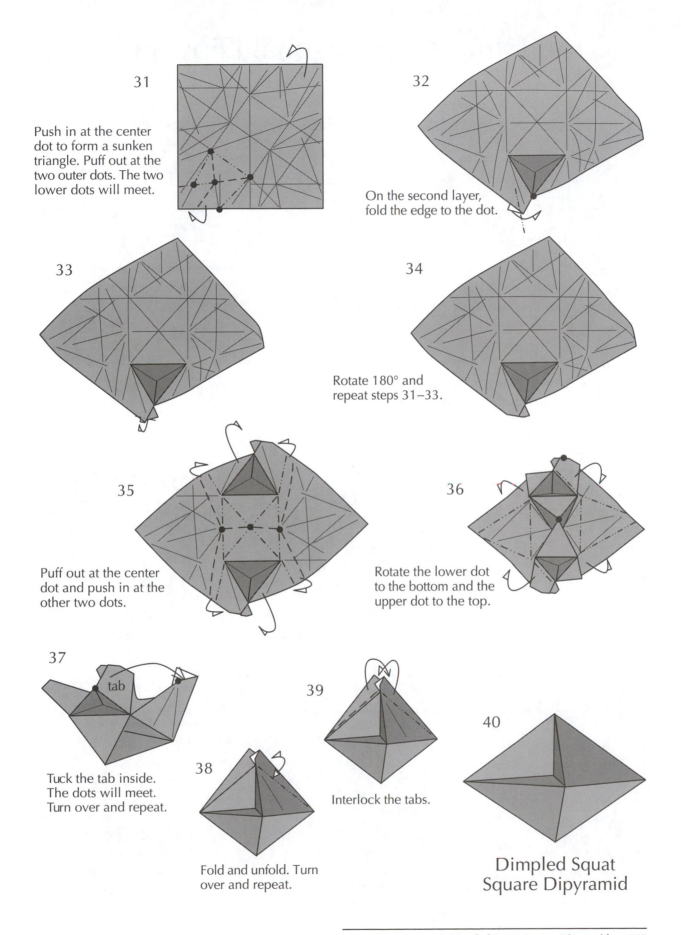

31

Push in at the center dot to form a sunken triangle. Puff out at the two outer dots. The two lower dots will meet.

32

On the second layer, fold the edge to the dot.

33

34

Rotate 180° and repeat steps 31–33.

35

Puff out at the center dot and push in at the other two dots.

36

Rotate the lower dot to the bottom and the upper dot to the top.

37

tab

Tuck the tab inside. The dots will meet. Turn over and repeat.

38

Fold and unfold. Turn over and repeat.

39

Interlock the tabs.

40

Dimpled Squat
Square Dipyramid

Tall Dimpled Hexagonal Dipyramid

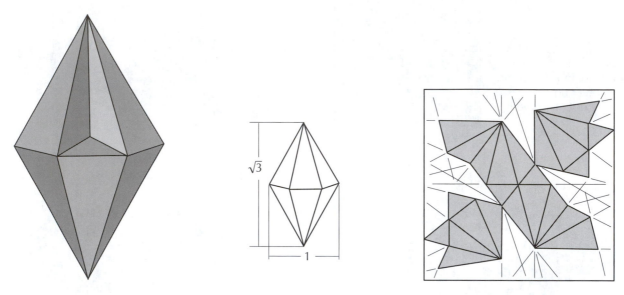

The ratio of the height of this dimpled dipyramid to its diameter is $\sqrt{3}$ to 1. The small angle in each nonsunken face is about 29° to achieve the dimensions.

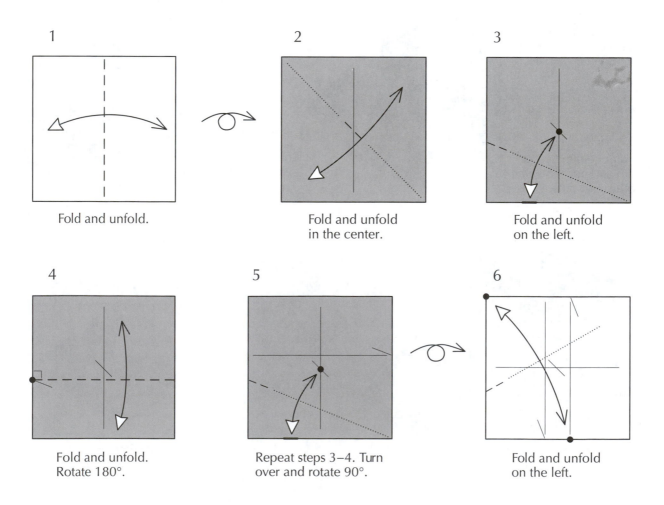

1

Fold and unfold.

2

Fold and unfold
in the center.

3

Fold and unfold
on the left.

4

Fold and unfold.
Rotate 180°.

5

Repeat steps 3–4. Turn
over and rotate 90°.

6

Fold and unfold
on the left.

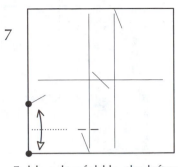

7

Fold and unfold by the leftmost vertical line. Rotate 180°.

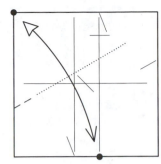

8

Repeat steps 6–7.

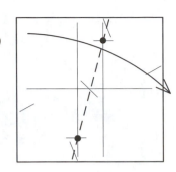

9

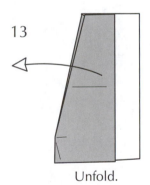

10

Valley-fold along the crease and mountain-fold the dot to the edge. Crease below the center line and a little above it.

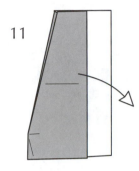

11

Unfold.

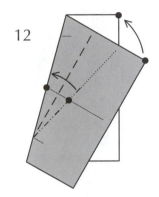

12

Repeat step 10.

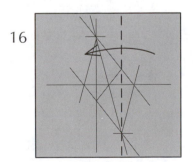

13

Unfold.

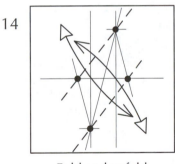

14

Fold and unfold.

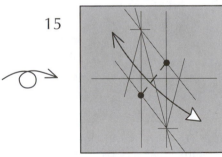

15

Fold and unfold.

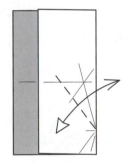

16

Fold along the crease.

17

Fold and unfold the top layer along a hidden crease.

18

Unfold.

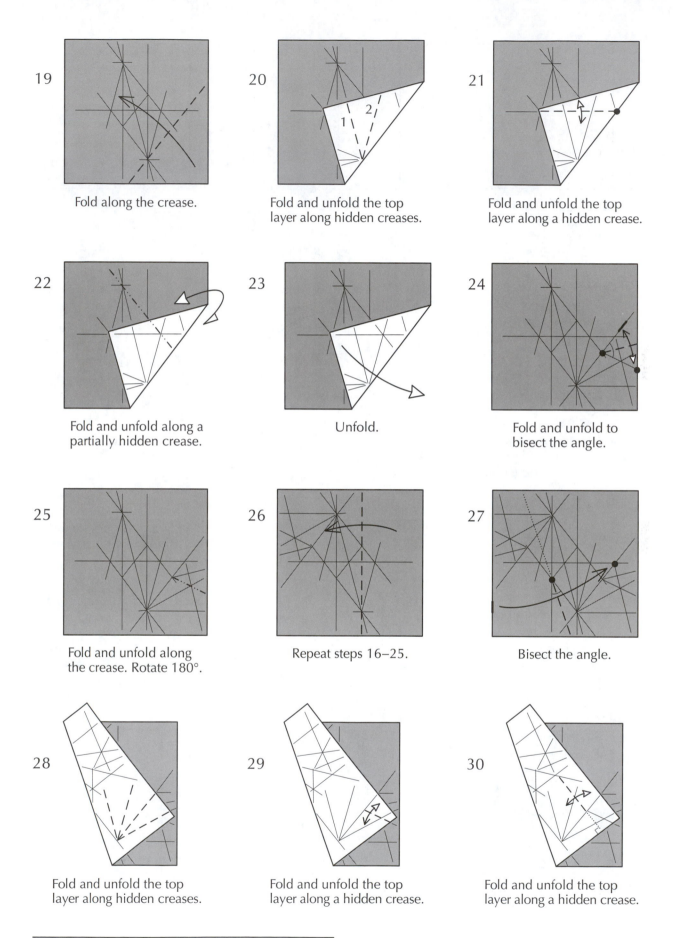

19   Fold along the crease.

20   Fold and unfold the top layer along hidden creases.

21   Fold and unfold the top layer along a hidden crease.

22   Fold and unfold along a partially hidden crease.

23   Unfold.

24   Fold and unfold to bisect the angle.

25   Fold and unfold along the crease. Rotate 180°.

26   Repeat steps 16–25.

27   Bisect the angle.

28   Fold and unfold the top layer along hidden creases.

29   Fold and unfold the top layer along a hidden crease.

30   Fold and unfold the top layer along a hidden crease.

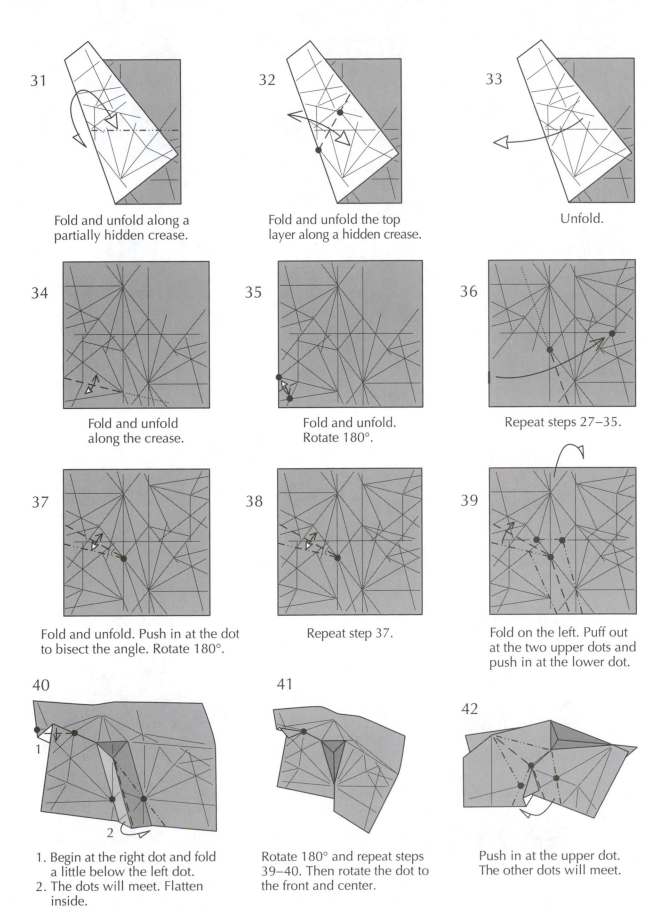

31

Fold and unfold along a
partially hidden crease.

32

Fold and unfold the top
layer along a hidden crease.

33

Unfold.

34

Fold and unfold
along the crease.

35

Fold and unfold.
Rotate 180°.

36

Repeat steps 27–35.

37

Fold and unfold. Push in at the dot
to bisect the angle. Rotate 180°.

38

Repeat step 37.

39

Fold on the left. Puff out
at the two upper dots and
push in at the lower dot.

40

1. Begin at the right dot and fold
   a little below the left dot.
2. The dots will meet. Flatten
   inside.

41

Rotate 180° and repeat steps
39–40. Then rotate the dot to
the front and center.

42

Push in at the upper dot.
The other dots will meet.

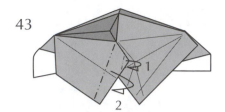

43

44

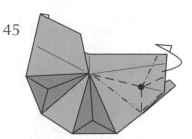

45

1. Make a small fold.
2. Fold along the crease.

Turn over and repeat steps
42–43. Then rotate the bottom
to the top and bring the dot to
the front and center.

Push in at the dot to
form a sunken triangle.

46

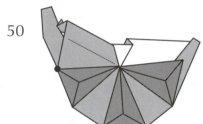

47

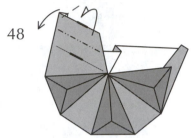

48

Squash-fold on the left and
bring the top edge to the crease.

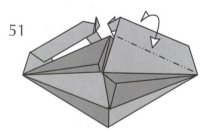

49

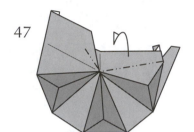

50

Turn over and repeat steps
45–49. Then rotate the dot
to the front and center.

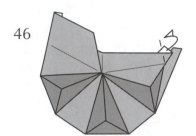

51

Fold and unfold. Turn
over and repeat.

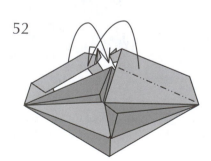

52

Tuck and interlock the tabs.

53

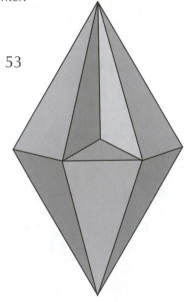

Tall Dimpled
Hexagonal Dipyramid

Dimpled Silver Hexagonal Dipyramid

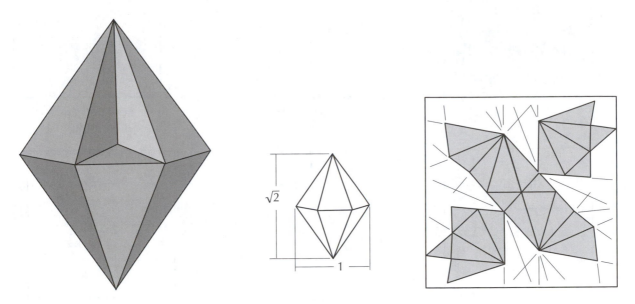

The ratio of the height of this dimpled dipyramid to its diameter is $\sqrt{2}$ to 1. The small angle in each nonsunken face is about 34° to achieve the dimensions. The folding is similar to that of the Tall Dimpled Hexagonal Dipyramid.

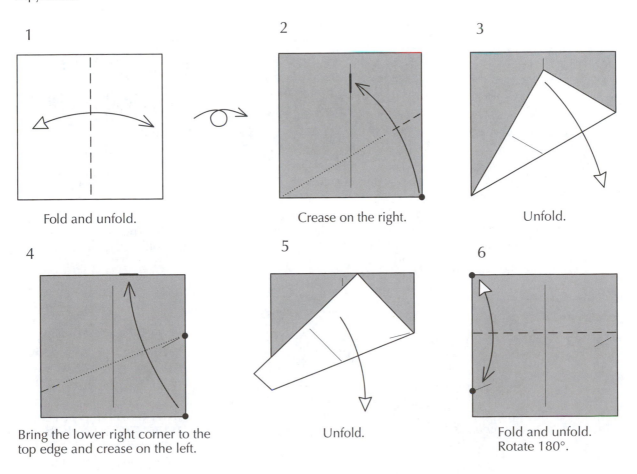

1

Fold and unfold.

2

Crease on the right.

3

Unfold.

4

Bring the lower right corner to the top edge and crease on the left.

5

Unfold.

6

Fold and unfold.
Rotate 180°.

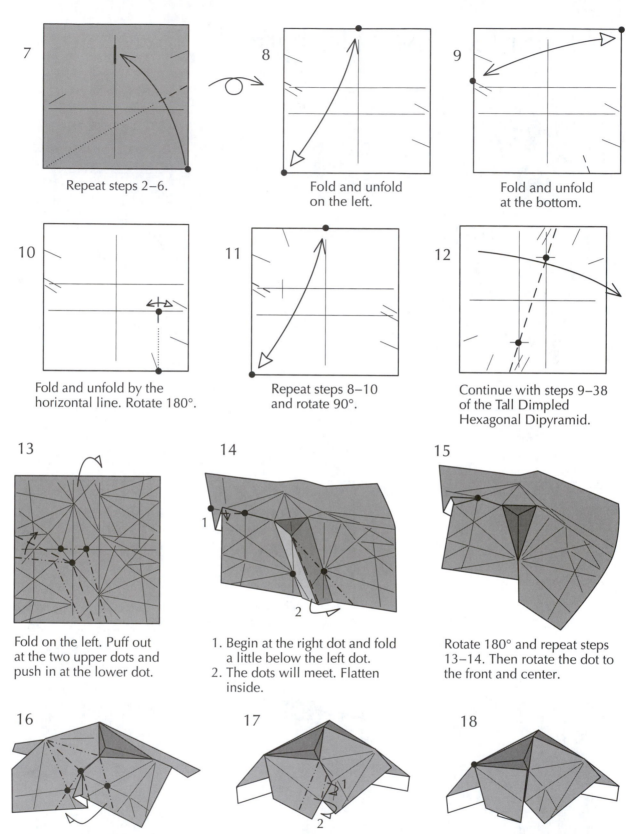

7

Repeat steps 2–6.

8

Fold and unfold
on the left.

9

Fold and unfold
at the bottom.

10

Fold and unfold by the
horizontal line. Rotate 180°.

11

Repeat steps 8–10
and rotate 90°.

12

Continue with steps 9–38
of the Tall Dimpled
Hexagonal Dipyramid.

13

Fold on the left. Puff out
at the two upper dots and
push in at the lower dot.

14

1. Begin at the right dot and fold
   a little below the left dot.
2. The dots will meet. Flatten
   inside.

15

Rotate 180° and repeat steps
13–14. Then rotate the dot to
the front and center.

16

Push in at the upper dot.
The other dots will meet.

17

1. Make a small fold.
2. Fold along the crease.

18

Turn over and repeat steps
16–17. Then rotate the bottom
to the top and bring the dot
to the front and center.

19

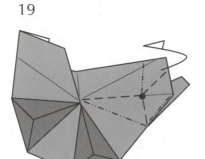

Push in at the dot to
form a sunken triangle.

20

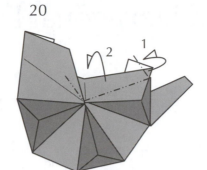

21

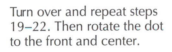

Squash-fold on the left so
the edge meets the dot.

22

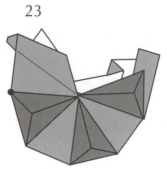

23

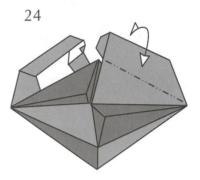

Turn over and repeat steps
19–22. Then rotate the dot
to the front and center.

24

Fold and unfold. Turn
over and repeat.

25

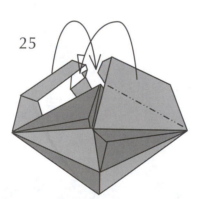

Tuck and interlock the tabs.

26

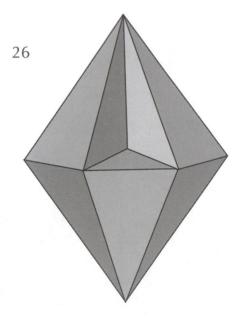

Dimpled Silver
Hexagonal Dipyramid

Dimpled Hexagonal Dipyramid

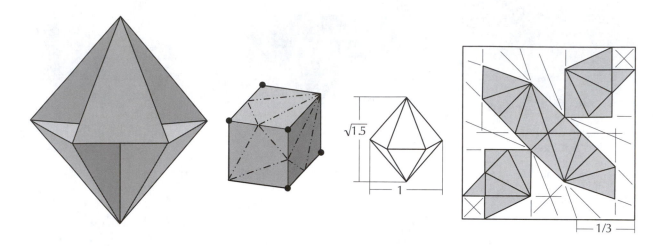

This model comes from a cube where six of the eight corners are sunken, as shown in the second picture. However, it was difficult to design a convenient model by folding a cube first. The angles in the nonsunken triangles are 36.87°, 71.565°, and 71.565°, and the ratio of the height to diameter is $\sqrt{1.5} = 1.225$. The folding is similar to that of the Tall Dimpled Hexagonal Dipyramid.

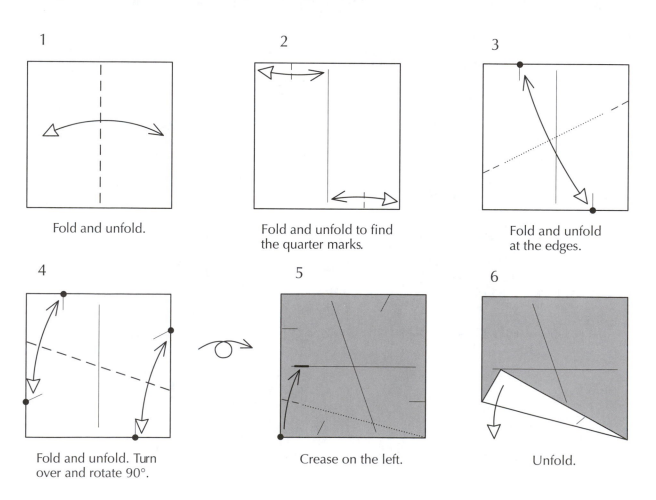

1

Fold and unfold.

2

Fold and unfold to find the quarter marks.

3

Fold and unfold at the edges.

4

Fold and unfold. Turn over and rotate 90°.

5

Crease on the left.

6

Unfold.

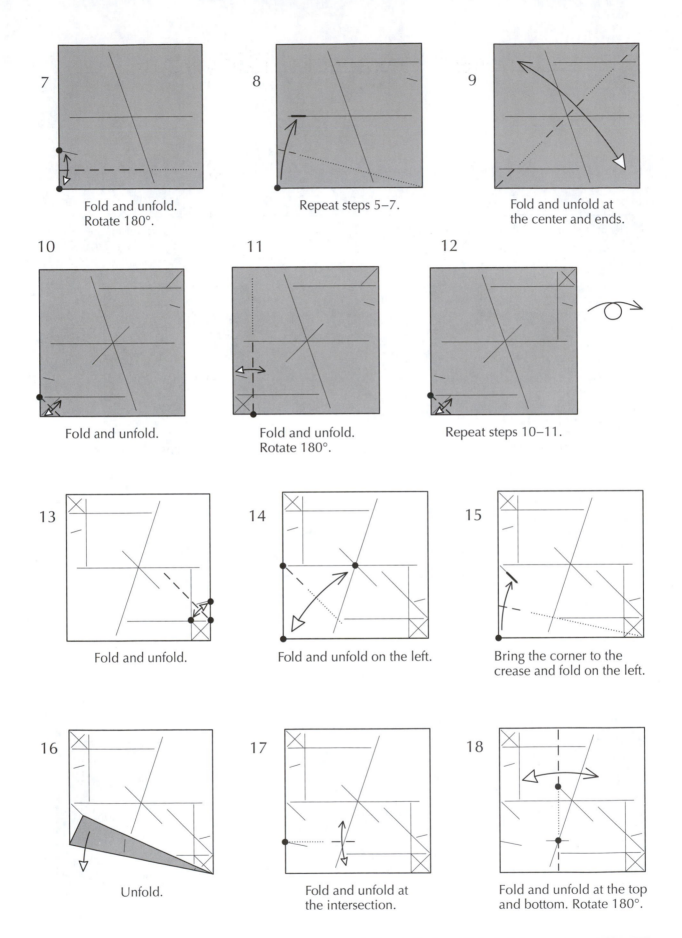

7 Fold and unfold.
Rotate 180°.

8 Repeat steps 5–7.

9 Fold and unfold at
the center and ends.

10 Fold and unfold.

11 Fold and unfold.
Rotate 180°.

12 Repeat steps 10–11.

13 Fold and unfold.

14 Fold and unfold on the left.

15 Bring the corner to the
crease and fold on the left.

16 Unfold.

17 Fold and unfold at
the intersection.

18 Fold and unfold at the top
and bottom. Rotate 180°.

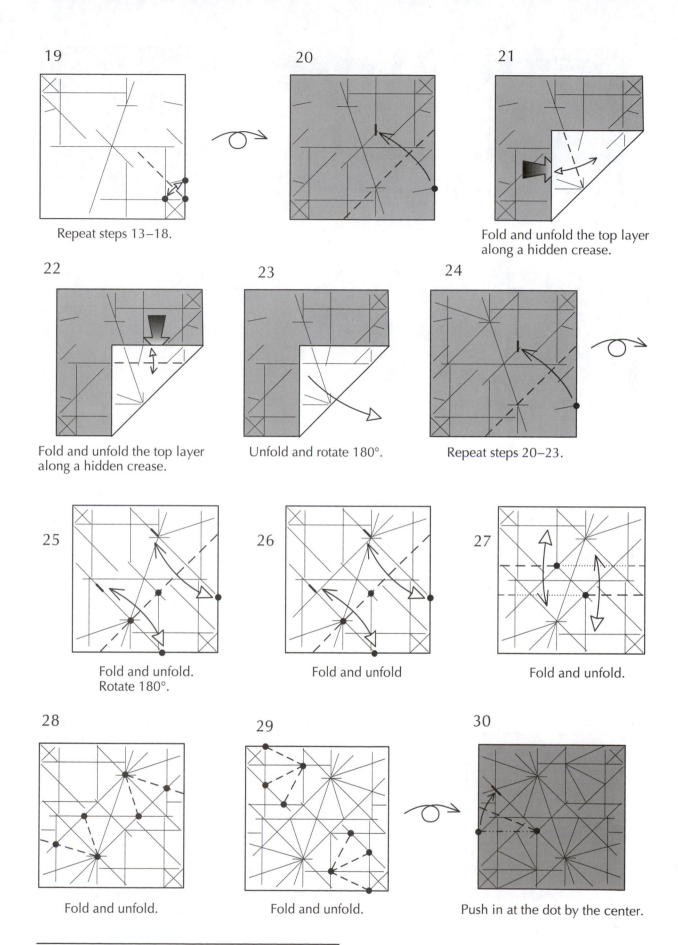

19

Repeat steps 13–18.

20

21

Fold and unfold the top layer
along a hidden crease.

22

Fold and unfold the top layer
along a hidden crease.

23

Unfold and rotate 180°.

24

Repeat steps 20–23.

25

Fold and unfold.
Rotate 180°.

26

Fold and unfold

27

Fold and unfold.

28

Fold and unfold.

29

Fold and unfold.

30

Push in at the dot by the center.

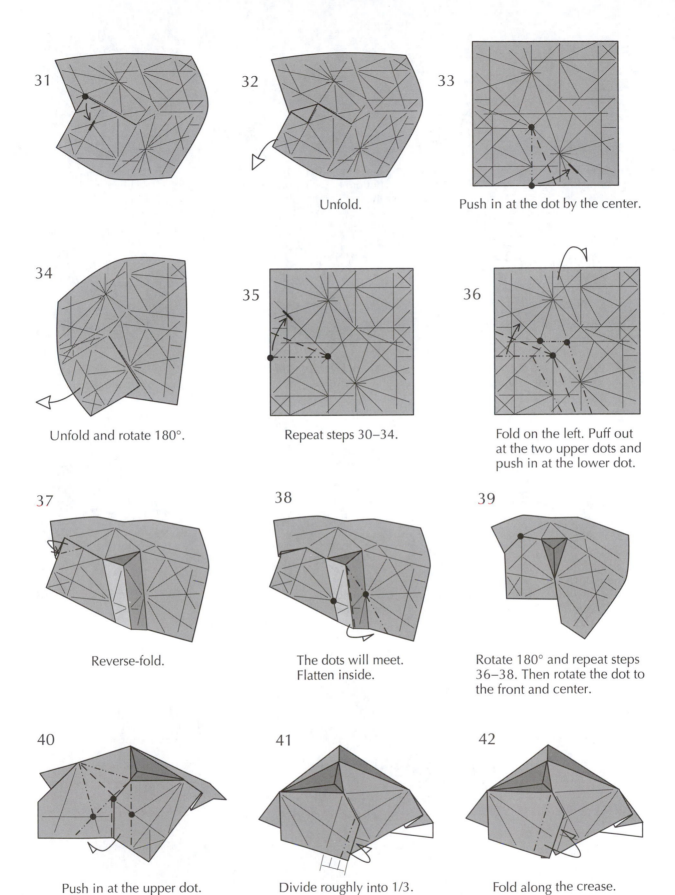

31

32

Unfold.

33

Push in at the dot by the center.

34

Unfold and rotate 180°.

35

Repeat steps 30–34.

36

Fold on the left. Puff out
at the two upper dots and
push in at the lower dot.

37

Reverse-fold.

38

The dots will meet.
Flatten inside.

39

Rotate 180° and repeat steps
36–38. Then rotate the dot to
the front and center.

40

Push in at the upper dot.
The other dots will meet.

41

Divide roughly into 1/3.

42

Fold along the crease.

43

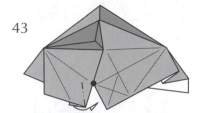

On the second layer,
bring the edge to the dot.

44

Turn over and repeat steps 40–43. Then
rotate the bottom to the top and bring
the dot to the front and center.

45

Fold along the creases.
Turn over and repeat.

46

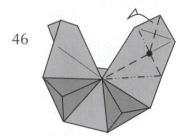

Push in at the dot to
form a sunken triangle.

47

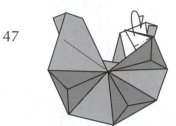

Fold both layers together
along the crease.

48

Tuck the white tab
inside the pocket.

49

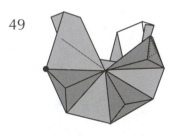

The hidden tab is shown with
the dotted lines. Turn over and
repeat steps 46–48. Then rotate
the dot to the front and center.

50

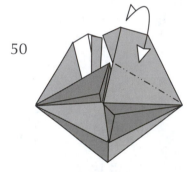

Fold and unfold. Turn
over and repeat.

51

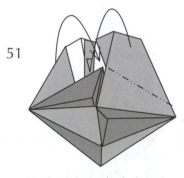

Tuck and interlock the tabs.

52

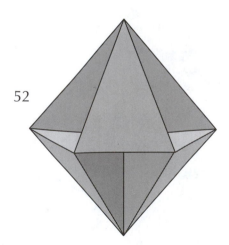

Dimpled Hexagonal Dipyramid

Dimpled Hexagonal Dipyramid in a Sphere

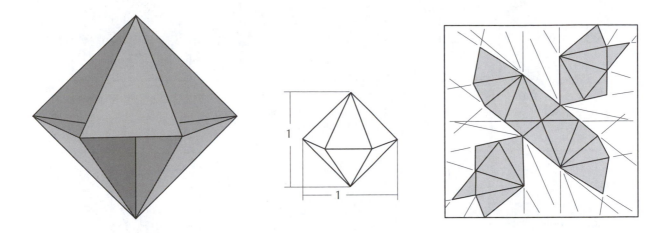

This dimpled dipyramid is inscibed in a sphere. The angles of each of the nonsunken triangles are 41.41°, 69.3°, and 69.3°. The folding is similar to that of the Tall Dimpled Hexagonal Dipyramid.

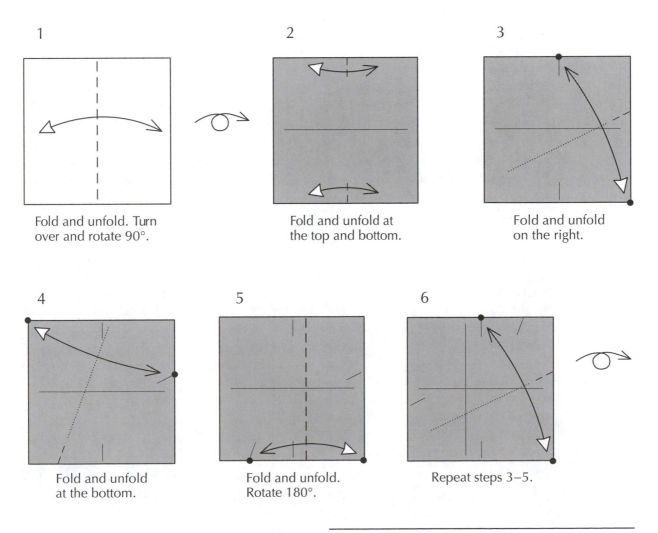

1

Fold and unfold. Turn over and rotate 90°.

2

Fold and unfold at the top and bottom.

3

Fold and unfold on the right.

4

Fold and unfold at the bottom.

5

Fold and unfold. Rotate 180°.

6

Repeat steps 3–5.

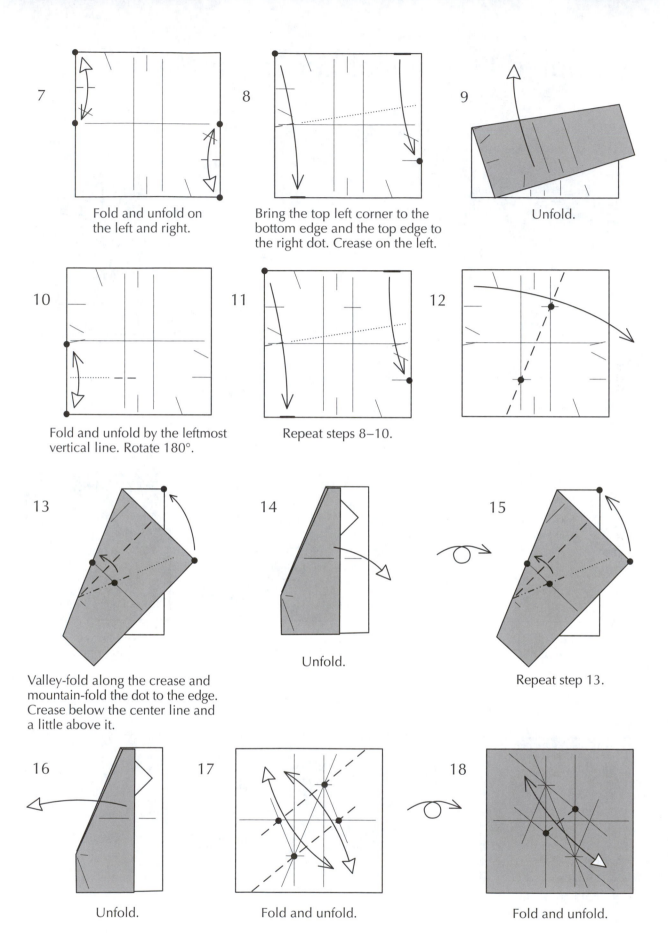

7 Fold and unfold on the left and right.

8 Bring the top left corner to the bottom edge and the top edge to the right dot. Crease on the left.

9 Unfold.

10 Fold and unfold by the leftmost vertical line. Rotate 180°.

11 Repeat steps 8–10.

12

13 Valley-fold along the crease and mountain-fold the dot to the edge. Crease below the center line and a little above it.

14 Unfold.

15 Repeat step 13.

16 Unfold.

17 Fold and unfold.

18 Fold and unfold.

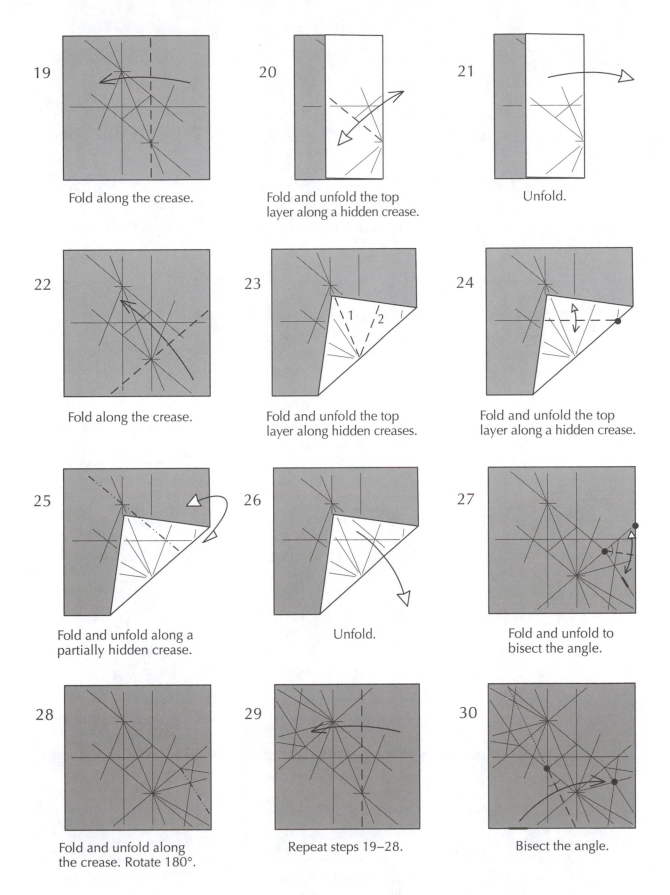

19 Fold along the crease.

20 Fold and unfold the top layer along a hidden crease.

21 Unfold.

22 Fold along the crease.

23 Fold and unfold the top layer along hidden creases.

24 Fold and unfold the top layer along a hidden crease.

25 Fold and unfold along a partially hidden crease.

26 Unfold.

27 Fold and unfold to bisect the angle.

28 Fold and unfold along the crease. Rotate 180°.

29 Repeat steps 19–28.

30 Bisect the angle.

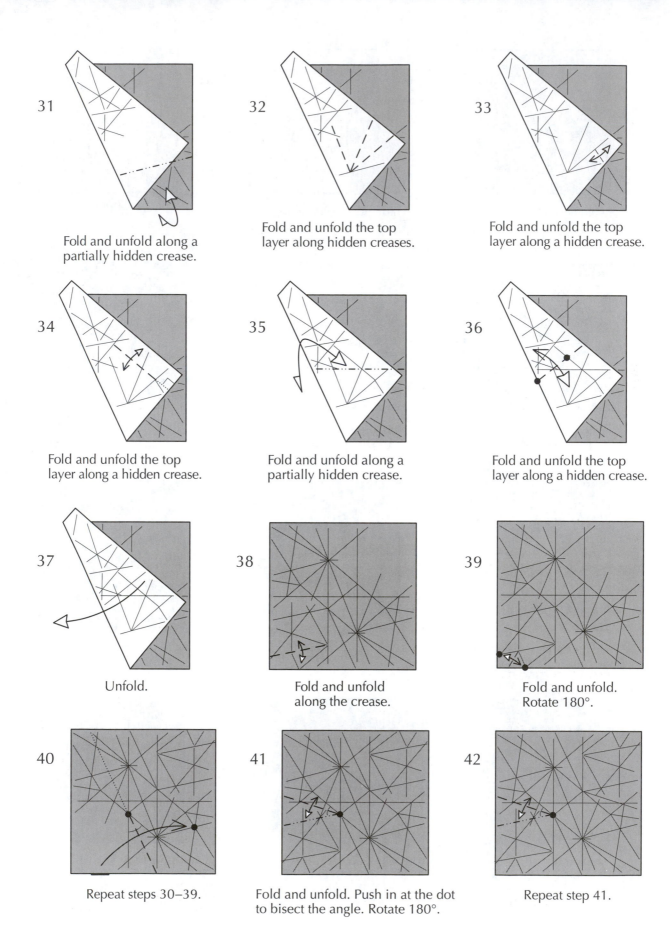

31 Fold and unfold along a partially hidden crease.

32 Fold and unfold the top layer along hidden creases.

33 Fold and unfold the top layer along a hidden crease.

34 Fold and unfold the top layer along a hidden crease.

35 Fold and unfold along a partially hidden crease.

36 Fold and unfold the top layer along a hidden crease.

37 Unfold.

38 Fold and unfold along the crease.

39 Fold and unfold. Rotate 180°.

40 Repeat steps 30–39.

41 Fold and unfold. Push in at the dot to bisect the angle. Rotate 180°.

42 Repeat step 41.

43

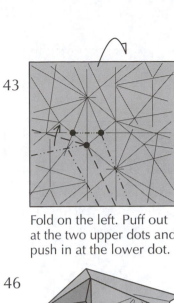

Fold on the left. Puff out at the two upper dots and push in at the lower dot.

44

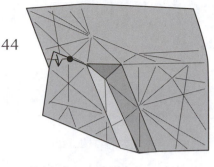

Fold along a hidden crease. Rotate the dot to the front and center.

45

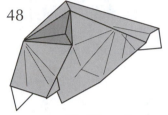

Push in at the upper dot. The other dots will meet.

46

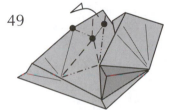

Mountain-fold along the crease and tuck between the layers.

47

48

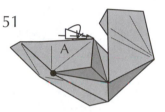

Repeat steps 43–47 on the back. Then rotate the bottom to the top.

49

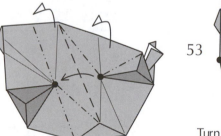

Push in at the lower dot. The other dots will meet.

50

51

Fold a thin strip and hide it behind region A. Rotate the dot to the front and center.

52

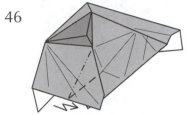

The dots will meet.

53

Turn over and repeat steps 49–52. Then rotate the dot to the front and center.

54

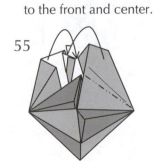

Fold and unfold. Turn over and repeat.

55

Tuck and interlock the tabs.

56

Dimpled Hexagonal Dipyramid in a Sphere

Octagonal Flying Saucer

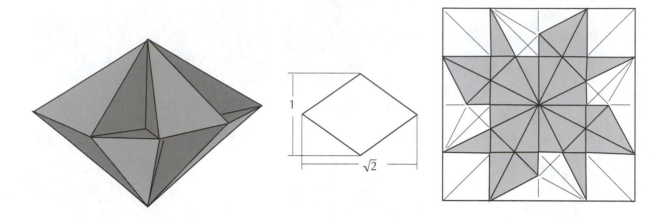

This model has an octagonal base. The ratio of the height to the diameter is $1/\sqrt{2}$. The crease pattern shows square symmetry.

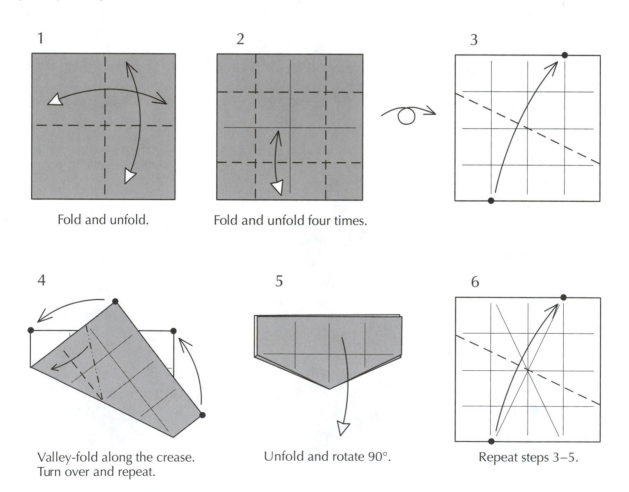

1

Fold and unfold.

2

Fold and unfold four times.

3

4

Valley-fold along the crease.
Turn over and repeat.

5

Unfold and rotate 90°.

6

Repeat steps 3–5.

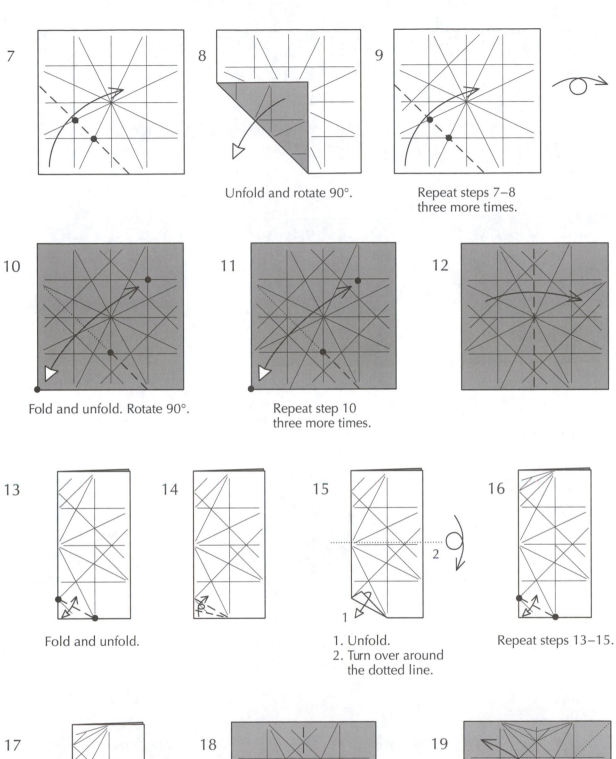

7

8

Unfold and rotate 90°.

9

Repeat steps 7–8
three more times.

10

Fold and unfold. Rotate 90°.

11

Repeat step 10
three more times.

12

13

Fold and unfold.

14

15

1. Unfold.
2. Turn over around
   the dotted line.

16

Repeat steps 13–15.

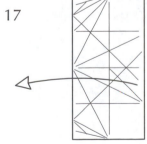

17

Unfold and rotate 90°.

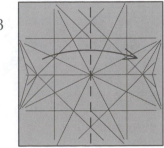

18

Repeat steps 12–17.

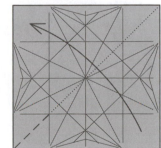

19

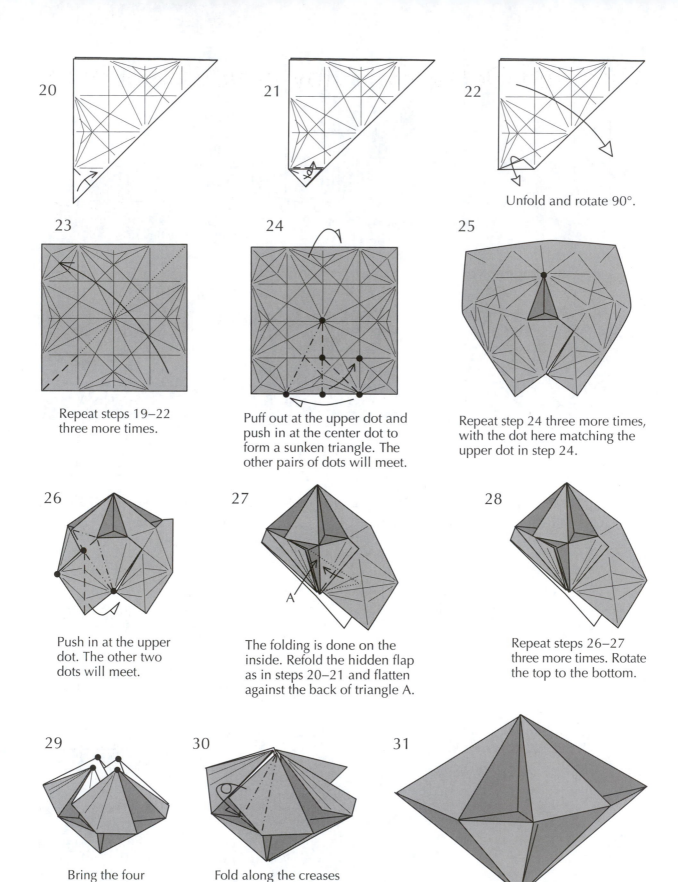

20

21

22

Unfold and rotate 90°.

23

Repeat steps 19–22
three more times.

24

Puff out at the upper dot and
push in at the center dot to
form a sunken triangle. The
other pairs of dots will meet.

25

Repeat step 24 three more times,
with the dot here matching the
upper dot in step 24.

26

Push in at the upper
dot. The other two
dots will meet.

27

A

The folding is done on the
inside. Refold the hidden flap
as in steps 20–21 and flatten
against the back of triangle A.

28

Repeat steps 26–27
three more times. Rotate
the top to the bottom.

29

Bring the four
dots together.

30

Fold along the creases
and tuck inside. Repeat
three more times.

31

Octagonal Flying Saucer

Dimpled Octagonal Dipyramid in a Sphere

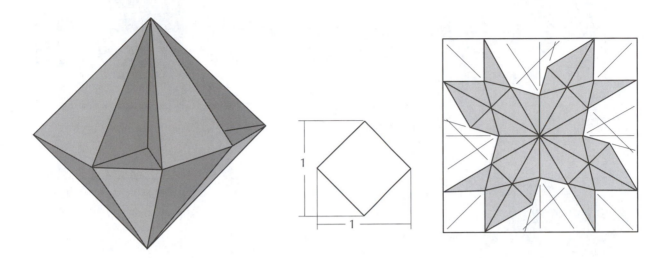

This dimpled dipyramid is inscibed in a sphere. The angles of each of the eight (nonsunken) triangles are 31.4°, 74.3°, and 74.3°.

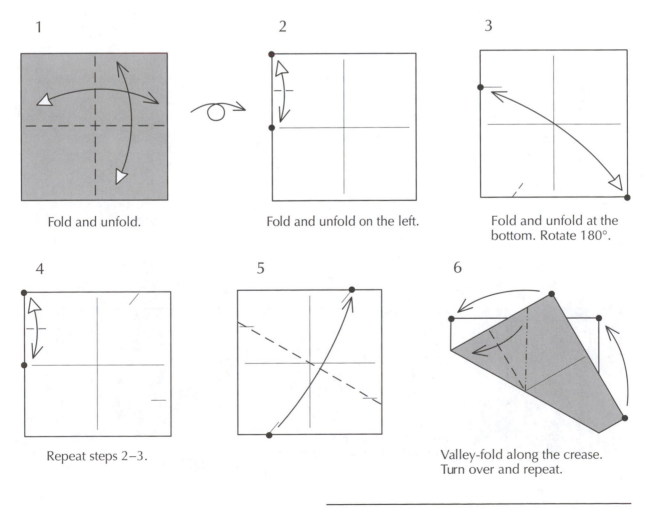

1

Fold and unfold.

2

Fold and unfold on the left.

3

Fold and unfold at the bottom. Rotate 180°.

4

Repeat steps 2–3.

5

6

Valley-fold along the crease. Turn over and repeat.

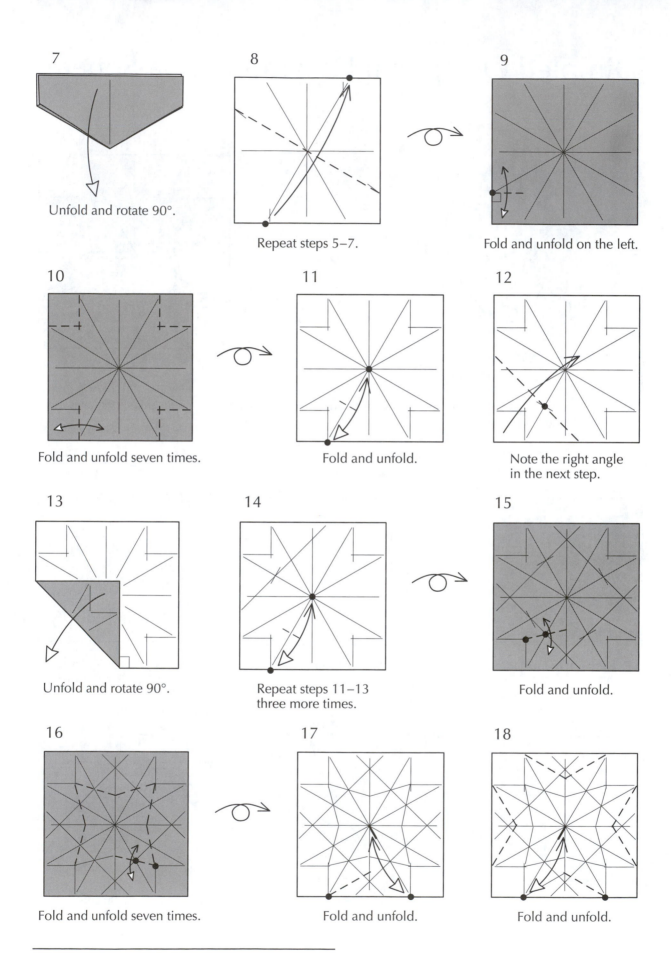

7

Unfold and rotate 90°.

8

Repeat steps 5–7.

9

Fold and unfold on the left.

10

Fold and unfold seven times.

11

Fold and unfold.

12

Note the right angle in the next step.

13

Unfold and rotate 90°.

14

Repeat steps 11–13 three more times.

15

Fold and unfold.

16

Fold and unfold seven times.

17

Fold and unfold.

18

Fold and unfold.

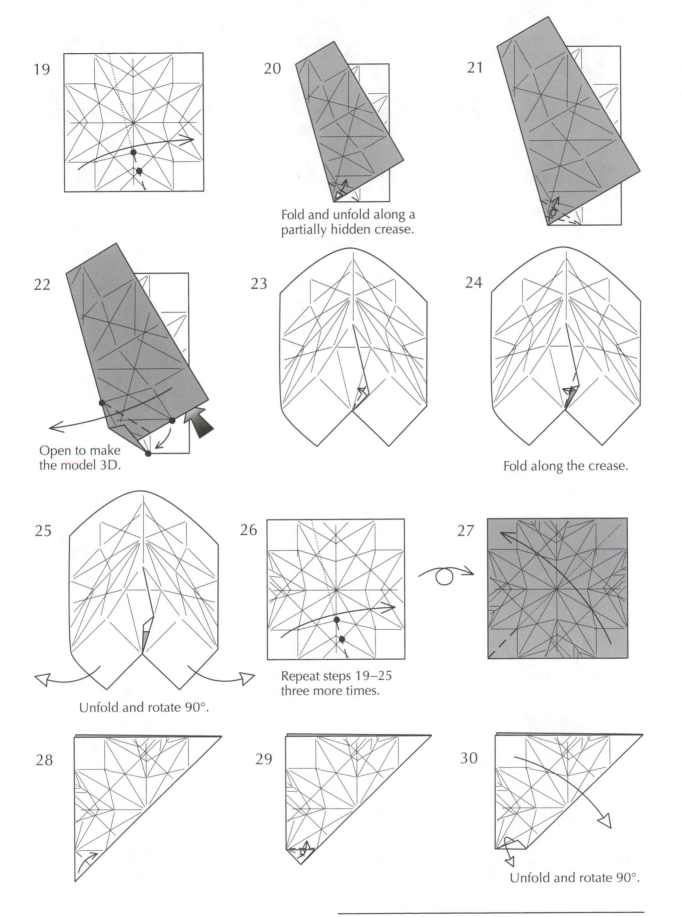

19

20

Fold and unfold along a
partially hidden crease.

21

22

Open to make
the model 3D.

23

24

Fold along the crease.

25

Unfold and rotate 90°.

26

Repeat steps 19–25
three more times.

27

28

29

30

Unfold and rotate 90°.

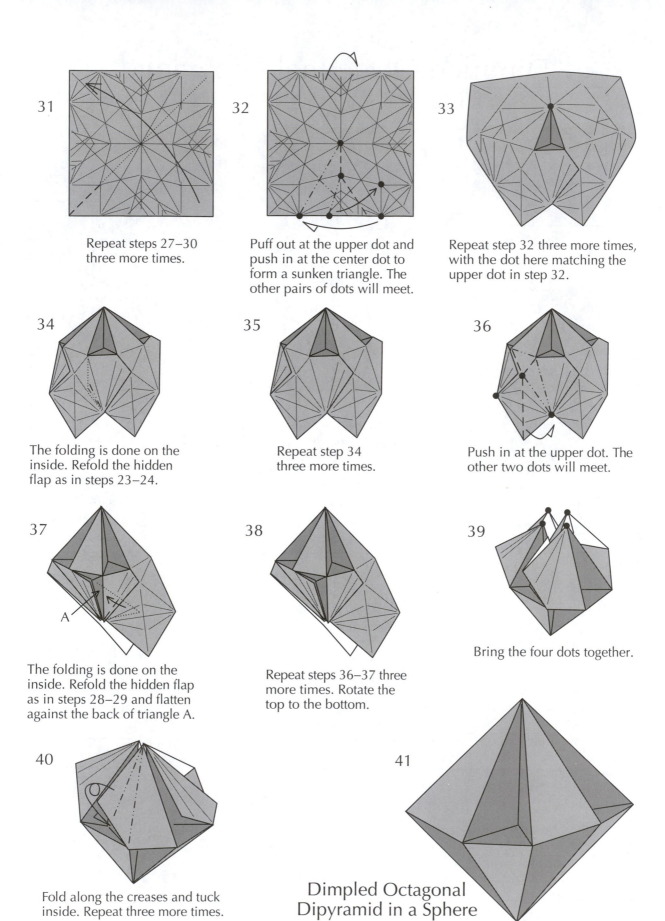

31 Repeat steps 27–30 three more times.

32 Puff out at the upper dot and push in at the center dot to form a sunken triangle. The other pairs of dots will meet.

33 Repeat step 32 three more times, with the dot here matching the upper dot in step 32.

34 The folding is done on the inside. Refold the hidden flap as in steps 23–24.

35 Repeat step 34 three more times.

36 Push in at the upper dot. The other two dots will meet.

37 The folding is done on the inside. Refold the hidden flap as in steps 28–29 and flatten against the back of triangle A.

38 Repeat steps 36–37 three more times. Rotate the top to the bottom.

39 Bring the four dots together.

40 Fold along the creases and tuck inside. Repeat three more times.

41 Dimpled Octagonal Dipyramid in a Sphere

Dimpled Octagonal Dipyramid

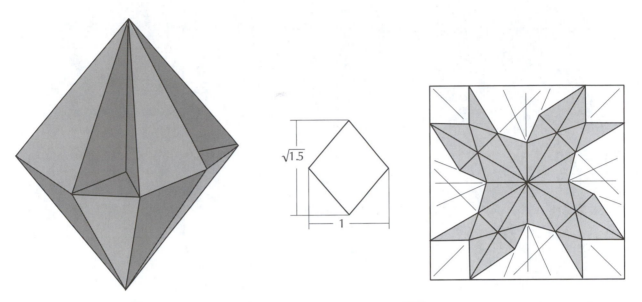

The ratio of the height of this dimpled dipyramid to its diameter is $\sqrt{1.5}$ to 1. The small angle in each nonsunken triangle is about 29° to achieve the dimensions. The folding is similar to that of the Dimpled Octagonal Dipyramid in a Sphere.

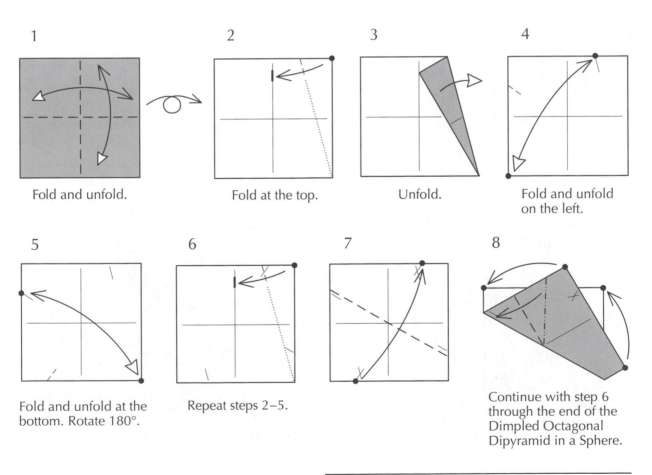

1

Fold and unfold.

2

Fold at the top.

3

Unfold.

4

Fold and unfold on the left.

5

Fold and unfold at the bottom. Rotate 180°.

6

Repeat steps 2–5.

7

8

Continue with step 6 through the end of the Dimpled Octagonal Dipyramid in a Sphere.

Dimpled Silver Octagonal Dipyramid

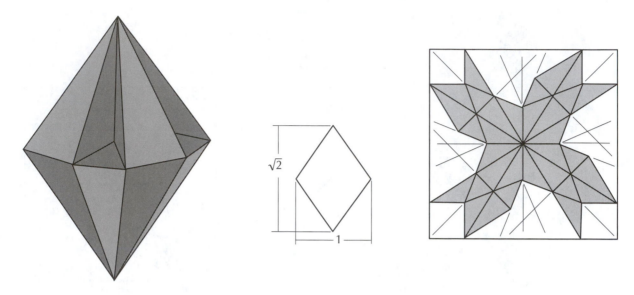

The ratio of the height of this dimpled dipyramid to its diameter is $\sqrt{2}$ to 1. The small angle in each nonsunken triangle is about 25.5° to achieve the dimensions. The folding is similar to that of the Dimpled Octagonal Dipyramid in a Sphere.

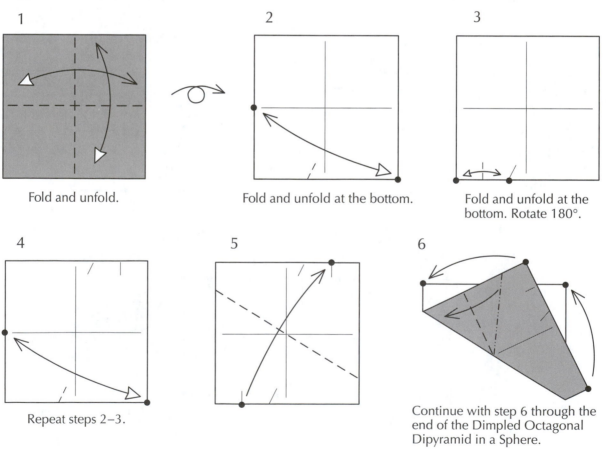

1

Fold and unfold.

2

Fold and unfold at the bottom.

3

Fold and unfold at the bottom. Rotate 180°.

4

Repeat steps 2–3.

5

6

Continue with step 6 through the end of the Dimpled Octagonal Dipyramid in a Sphere.